# WISCONSIN
# In Bloom

## Find and Identify the 141 Most Common Wildflowers

Michael Homoya

TIMBER PRESS
PORTLAND, OREGON

Timber Press
Workman Publishing
Hachette Book Group, Inc.
1290 Avenue of the Americas
New York, New York 10104

timberpress.com

Timber Press is an imprint of Workman Publishing, a division of Hachette Book Group, Inc. The Timber Press name and logo are registered trademarks of Hachette Book Group, Inc.

Printed in Dongguan, China, (TLF) on responsibly sourced paper

Text layout by Mary Velgos, based on series design by Hillary Caudle
Cover design by Leigh Kaisen

Endpaper illustration by Alan Bryan

The publisher is not responsible for websites (or their content) that are not owned by the publisher.

ISBN 978-1-64326-458-5

A catalog record for this book is available from the Library of Congress.

In memory of my brothers Peter and Bruce

# Contents

Consider the lilies of
the field, how they
grow; they toil not,
neither do they spin:
And yet I say unto
you, that even
Solomon in all his
glory was not arrayed
like one of these.
—Matthew 6:28–29

# PREFACE



**Wisconsin is a wonderland of wildflowers!** It consists of a diverse "mash-up" of North Woods and wetland plants with those of the tallgrass prairie and the profusely rich, verdant hardwood forests of the East. More than 1800 native species of vascular plants have been recorded here, and about 500 additional species have been introduced and naturalized from elsewhere in the world. There are wildflowers of every size, shape, and color, including nearly 50 species of native orchids, such as the "pretty-in-pink" tuberous grass pink orchid (*Calopogon tuberosus*) and the strikingly beautiful showy lady's slipper (*Cypripedium reginae*).

Wildflowers are most common in natural areas, but even if you have never set foot in such a place, you may already know, or at least be familiar with, some of these plants, such as wild species of iris, geranium, and lily. Of course, others may not be so familiar, such as the rare kittentails (*Besseya bullii*) and glade mallow (*Napaea dioica*). Bizarre plants are here too. Consider dodder (*Cuscuta* species), a leafless flowering parasite that looks a bit like yellow-orange strands of cooked spaghetti, and the carnivorous purple pitcher plant (*Sarracenia purpurea*), with vaselike leaves that contain deadly pools of liquid that trap and digest helpless insects. Whether they are colorful and showy or bizarre and deadly, these plants and more are presented in this easy-to-use field guide that features photographs and descriptions of some of Wisconsin's more common and interesting wildflowers. (For a treatment covering many additional wildflower species found in the state, consider *Wildflowers of the Midwest* by Homoya and Namestnik.)

Why learn about wildflowers? One incentive is that they are good indicators of natural quality, and the higher the quality, the better the health of the land (and you too!). If for no other reason, learning to identify plants (and animals) helps you "avoid or overcome the monotony of anonymous scenery," as my botany professor Dr. Robert Mohlenbrock wrote so aptly years ago. It is my hope that this book not only will help make Wisconsin's natural land less anonymous to you, but will also entice you to cherish and protect it. Wisconsin writer and naturalist Aldo Leopold wrote, "When we see land [and the associated wildflowers] as a community to which we belong, we may begin to use it with love and respect."

# ACKNOWLEDGEMENTS

**No book is ever created by the efforts** of only one person, and that is certainly the case with this one. I am greatly indebted to the many people who helped me in so many ways, from tracking down information to reviewing the draft. For all their help, I wish to thank Michele Angel, Harvey Ballard, Donna Baron, David Boufford, Francis Bremer, Teresa Clark, Moira Fitzgerald, Ryan Harrington, Daniel Hinchen, Jacoba Lawson, Adrienne Mayor, Will McKay, Sarah Milhollin, Julia Morrow, Jessica Murphy, Scott Namestnik, James Pringle, Andrea Rapacz, Tony Reznicek, Pat Schaefer, Jeannie Sherman, Reiner Smolinski, Lisa Theobald, and Gerould Wilhelm. It is much appreciated.

Of course, this book wouldn't be what it is without the photographs. While some are mine, most were captured by others, and for the use of them I offer my sincerest thank you to Adam Balzer, Katy Chayka, Peter Dzuik, Andrew Gibson, Peter Grube, Karen Hoksbergen for use of the image taken by Dan Tenaglia, Michael Huft, Eric Hunt, Scott Namestnik, Nathanael Pilla, Corey Raimond, Paul Rothrock, Perry Scott, and Brad Slaughter.

My lovely wife, Barbara, helped me in many aspects of producing this book, including reading and editing several of the drafts and, perhaps most importantly, providing moral support (and patience) during the research and writing phases. Our two sons, Aaron and Wesley Homoya, were invaluable whenever I needed help with computer issues. Thank you all, dear family.

# Getting to Know Wisconsin

**Wisconsin is a land of natural delights**, ranging from the spectacular sandy and rocky shores along Lake Michigan and Lake Superior, to the rugged, unglaciated hills and rocky ravines of the Driftless Area in the southwest. Describing all the natural features within the state's 65,000 square miles of land and water would require an entire book; rather than doing that, I have simplified the discussion by dividing the state into two ecoregions: the Northern Lakes Ecoregion and the Tallgrass Prairie Ecoregion.

The Northern Lakes Ecoregion, which covers most of the northern half of the state, owes much of its appearance to glaciation. During the Pleistocene Epoch several millennia ago, the region was trounced by several advances of massive ice sheets that measured up to a mile thick in some places. The resulting landscape we are graced with today comprises a complex of forests, picture-perfect lakes, moraines, till plains, and lake plains. This ecoregion is also where we find the greatest occurrence of peatlands in the state, most in the form of marshes, swamps, bogs, and fens. These saturated peatlands are commonly quite open, but many are forested, with distinctive trees such as black ash, paper birch, yellow birch, white cedar, balsam fir, red maple, black spruce, and tamarack. On the better drained sites, we find bigtooth aspen, quaking aspen, American basswood, black cherry, eastern hemlock, sugar maple, northern red oak, and white pine. Jack pine and red pine are characteristic of drier sites.

Most of the southern part of the state is included within the Tallgrass Prairie Ecoregion. Open terrain in such natural communities as prairie, savanna, barrens, and woodlands reigns supreme here—or once did. Most of these natural

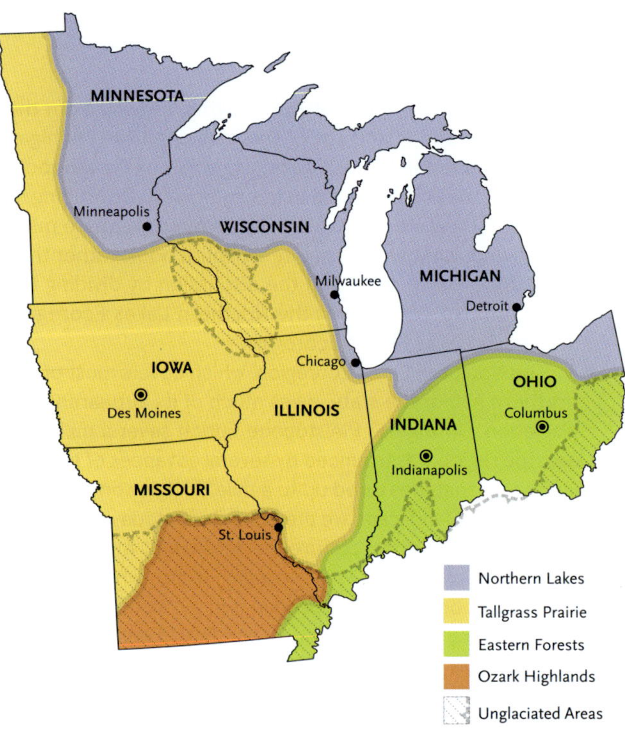


communities have been altered or eliminated by land conversion or fire suppression to the degree that only remnants remain. Forests are also present, scattered in patches, especially in areas along water courses or hillsides too steep to cultivate. They comprise mostly broadleaf deciduous species. In the drier sites are several oak species, such as black oak, bur oak, northern pin oak, northern red oak, and white oak. Other notable species, especially in moist sites, include white ash, black cherry, eastern cottonwood, American elm, slippery elm, hackberry, bitternut hickory, shagbark hickory, eastern hop hornbeam, sugar maple, and red mulberry.

Near the middle of the state is a "tension zone," where the two ecoregions meet. Simply put, plants and animals of northern affinities meet those of the south in this area. In this zone, the northern forests, dominated by coniferous needle-bearing trees, meet the southern and eastern forests, composed predominantly of broadleaf deciduous trees. Be aware that there is considerable overlap of species and natural communities between the two ecoregions—nature does not usually abide by the lines and boundaries we assign to it.

In the southwestern portion of the state is a special area known as the Driftless Area. It consists of land mostly untouched by continental glaciation and consequently is composed of rugged hills, valleys, and deep-cut, rocky gorges and cliffs. It provides a home for a variety of wildflowers, some of which are found in the state only here.

## Where to look for wildflowers

Fortunately, there is no shortage of places in Wisconsin to look for wildflowers. National and state forests, lakeshores, nature preserves, parks, scenic riverways, and wildlife refuges are all excellent places to explore. All combined, the state contains many thousands of acres of public lands. Private property offers even more areas—just be certain to acquire permission before visiting.

Several public properties are organized here by the ecoregion in which they occur. Readers should contact the owners or site managers for specific up-do-date information before visiting properties of interest. Some of these areas can be remote and may not have maintained trails or other recreational facilities.

**Tallgrass Prairie Ecoregion**

- Blue Mound State Park
- Perrot State Park
- Puchyan Prairie State Natural Area
- Rocky Run Oak Savanna State Natural Area
- Rush Creek State Natural Area
- Spring Green Preserve and State Natural Area
- Wyalusing State Park
- York Prairie State Natural Area

**Northern Lakes Ecoregion**

- Apostle Islands National Lakeshore
- Big Bay State Park
- Brule River State Forest
- Chequamegon-Nicolet National Forest
- Interstate State Park
- Newport State Park
- Northern Highland–American Legion State Forest
- Plover River Woods State Natural Area
- St. Croix National Scenic Riverway (shared with Minnesota)

## Natural Communities

Wildflowers are where you find them. But finding them doesn't need to be a random exercise. For the most part, plants occur in specific natural community types (habitats), where organisms live, interact, and obtain resources to survive. Thus, if you're looking for a particular wildflower, you'll find it advantageous to seek out and locate the natural community in which it is known to grow. Natural communities are often classified by dominant vegetation, soil type, moisture condition, and topographic position. The state of Wisconsin includes several general natural community types, which are referenced in this book as barrens, bogs, fens, forests, lakes, marshes, prairies, savannas, springs, streams, rivers, swamps, and woodlands, as well as disturbed sites created by humans.





**BARRENS** A usually dry, nutrient-poor site with sparse vegetation. If trees are present, they are often stunted and twisted. Barrens are often excessively drained areas with deep sand.

**BOG** A typically open and acidic, nutrient-poor wetland, commonly with a floating mat of living and decaying sphagnum moss dominated by heath family shrubs. Usually in a basin, not fed by groundwater.

**DISTURBED SITES** An area that has been disturbed mostly by human activity, such as agricultural fields, industrial sites, rail yards, ditches, roadsides, and wastelands. Plants growing in these sites are often invasive exotics, but some native species such as common milkweed (*Asclepias syriaca*) may thrive here too.

**FEN** A site with highly organic soil saturated with groundwater, usually flowing in a slow, diffuse manner. Fens can be alkaline, circumneutral (nearly neutral) or acidic, open with lots of sedges and forbs (nonwoody, nongrass-like flowering plants), or forested.

**FOREST** A site dominated by trees that usually have tall, straight trunks and commonly form a closed upper canopy. Layers of smaller trees and shrubs form the midstory and understory. Occurs on sites ranging from very wet (swamp) to dry.

**LAKE** A relatively large body of water that is often sufficiently deep that only aquatic plants can survive. Ponds have similar characteristics but are smaller in size.

**MARSH** A wetland dominated by sedges, grasses, cattails, and rushes. The water level is usually relatively shallow.

**PRAIRIE** A mostly treeless area dominated by native grasses and forbs.

**SAVANNA** An area dominated by grasses and forbs but interspersed with open-grown and widely spaced trees. Generally, less than 30 percent of the canopy cover is trees.

**SPRING** A natural, continuous or intermittent flow of water emerging from the ground.

**STREAM/RIVER** Flowing waters confined within channels and banks occurring in valleys and floodplains. Associated communities include gravel bars, sand bars, mudflats, and banks.

**SWAMP** A forested or shrubby wetland, usually with standing water, the source of which is direct precipitation, flooding, or groundwater seepage.

**WOODLAND** Similar to forest, but spacing of trees is typically less dense and typically do not form a closed canopy. Midstory and understory are usually sparse due to substrate conditions and/or fire.

# Plant Species and Names

**People have been naming organisms** since time began, but it was 17th-century English naturalist John Ray who first defined "species" as the basic level of classification and naming. Generally speaking, a species is a group of organisms that share physical and genetic characteristics and can interbreed and produce fertile offspring. Species can have two kinds of names: a common name and a scientific name. Common names can be interesting but are also messy, because many plants have more than one common name, and a problem arises when a plant is known by one name in one area and a different name elsewhere. For that reason, I highly recommend that you learn scientific names, which are usually Latin or Latinized words. They are standardized, and regardless of where in the world you live, the scientific name for a particular species is the same. It consists of two parts: genus and specific epithet. The combination of the words is known as a binomial. Don't be intimidated by the Latin. You may not realize it, but you are likely already familiar with this naming convention. For example, you probably know *Tyrannosaurus rex* and our own scientific name, *Homo sapiens*. Now consider one of our more common wildflowers in the state, the northern blue flag. *Iris* is the genus name and *versicolor* the specific epithet. Together they form the binomial scientific name *Iris versicolor*. Another native *Iris* in Wisconsin is the dwarf lake iris (*I. lacustris*). Because these two plants share the same genus name, we know that they are obviously related, but the different specific epithets tell us that the two are different species and to a degree are consistently different from each other.

## Plant Family Classifications

A plant family represents another level of classification that groups related plants together. Families are based on similarity of reproductive structures, typically the flower and fruit. Today, due to advances in technology, family designation is also determined by an organism's genetic makeup. Just like human families, plant families show relationships. Often it is apparent that a particular plant, even if it's initially unknown, belongs to a particular family. For example, one trait of members of the carrot family is that their flowers are commonly arranged in an umbel (in which short flower stalks spread from a common point, kind of like the ribs of an inverted umbrella). Although some plants with umbels belong to other families, identifying an unknown plant having an umbel should at least suggest that it *could* be a member of the carrot family.

### Plant family characteristics

The characteristics provided in these descriptions are focused on species belonging to the plant families treated here. A particular family's overall diversity, however, may display other characteristics or growth forms that are not mentioned for the Wisconsin species. For example, some violet species grow as shrubs (in Hawaii!) but do not do so in Wisconsin. In some of the accounts, house plants or plants of the garden or grocery produce section are mentioned to show relationships between our wildflowers with plants that may be more familiar to you.

**Sweetflag family (Acoraceae)**

This small family of herbaceous wetland plants features long, sword-shaped leaves with parallel veins, somewhat like those of iris, that produce a sweet odor when bruised. Flowers are inconspicuous and regular, with six tepals (sepals and petals indistinguishable), numerous on a cylindrical club (spadix). The species were formerly placed in the arum family (Araceae).





**Water plantain family (Alismataceae)**

These herbaceous perennial plants grow in water or mud. Leaves are usually simple and entire. Flowers have three white petals, arranged on a stalk with a whorled branching pattern on a terminal flowering stalk.

**Onion family (Alliaceae)**

These perennial herbaceous plants usually grow from bulbs, with simple narrow leaves (or broad in wild leek), mostly basal, sometimes alternate. Most have the characteristic onion or garlic scent when bruised. Flowers are regular with six tepals, usually in a rounded cluster atop the flowering stalk, sometimes replaced by bulblets.

**Carrot family (Apiaceae)**

Members of this family are usually easily recognized by their flat-topped or rounded umbel of flowers (with stalks spreading from a single point). Leaves are mostly alternate, often highly divided, and have sheathing bases that wrap around the stem. Flowers are small with five petals. In addition to our wild plants are several important food plants in the family, including carrot, parsley, cilantro, parsnip, dill, and others.

### Dogbane family (Apocynaceae)

These perennial herbaceous plants have simple, mostly opposite, entire leaves with milky sap. Flowers have five petals. Milkweed (*Asclepias* species) flowers have five tubular hoods, each with a horn, in clusters atop the plant or in leaf axils. Seedpods are swollen and lance- or narrowly egg-shaped, or they may resemble bean pods (in dogbanes, *Apocynum* species). Seeds often have silky hairs at the tips to help them disperse in wind. In Wisconsin, milkweeds outnumber dogbane species.

### Arum family (Araceae)

These herbaceous plants are characterized by having a spathe (hood) and a spadix (club) upon which inconspicuous flowers are attached. Flowers lack petals. The plants contain calcium oxalate crystals and thus are not to be eaten. This family is famous for its many tropical species, and several are used for their interesting foliage as house plants, such as heartleaf philodendron, calla lily, red heart anthurium, dumb cane, and pothos.

### Ginseng family (Araliaceae)

This family of perennial herbaceous plants and shrubs has mostly alternate, whorled, simple or compound leaves. Flowers are small and regular, with five petals in an umbel (the stalks spreading from a single point) or in a cluster of small umbels. The highly valued ginseng is harvested from the wild as well as being cultivated.

### Pipevine family (Aristolochiaceae)

The leaves of these perennial herbaceous plants are mostly heart- or kidney-shaped and entire. Flowers with three petal-like sepals form an unusual shape, somewhat like a miniature pipe or urn. Elsewhere, several members of the family are vines, such as Dutchman's pipevine that mostly occurs farther south.

### Aster family (Asteraceae)

The Asteraceae is considered one of the largest plant families in the world and includes thousands of species. It was originally known as the composite family (Compositae). Flowers are packed together on a specialized head that often results in an impression of a large, single flower (think sunflower). Two types of flowers are commonly present on the head, disk flowers and ray flowers, though many species' flowers feature just one type. Disk flowers are typically tubular in shape, tiny, and packed together on the disk in the center of the head. Ray flowers are much larger and often flat, and they encircle or completely cover the disk. The base of the head is covered with tiny leaflike bracts.

### Touch-me-not family (Balsaminaceae)

Our species of *Impatiens*, the local representatives of the touch-me-not family, are herbaceous annuals with juicy, fleshy stems and simple leaves. Flowers are irregular, usually with three petals and three sepals, the lower one with a nectar spur. The fruit is typically a capsule that, when ripe, explodes upon touching or other disturbance to expel its seeds. Exotic species are popular bedding plants, such as New Guinea impatiens.

**Barberry family (Berberidaceae)**
Native members of this family in Wisconsin are perennial herbs with simple or compound leaves. Flowers are regular, usually with six petals or petal-like sepals. The fruit is a berry (in mayapple) or a berry-like seed (in blue cohosh). Japanese barberry (*Berberis thunbergii*), a shrub, is commonly used in landscaping and is also quite invasive.

**Borage family (Boraginaceae)**
These plants, often bristly-hairy, have simple basal or alternate leaves with regular, somewhat tubular flowers with five lobes. They are commonly in coiled clusters that unfurl as blooming commences. The true forget-me-not (*Myosotis scorpioides*) is nonnative and quite invasive in several moist habitats.

**Mustard family (Brassicaceae)**
This large family consists of several annual, biennial, and perennial species that often have pungent, watery sap. Leaves are alternate and/or basal, simple or deeply pinnately lobed. Flowers are typically regular, with four mostly white or yellow petals. Fruit is a distinctive, mostly narrow capsule that, when mature, splits into two halves to expose the seeds. Also known as the Cruciferae, a reference to the crosslike arrangement of the petals.

**Watershield family (Cabombaceae)**
These perennial aquatic plants have alternate floating leaves. The leaf stalks are attached to the central portion of the blade. Flowers typically are regular with three petals, usually occurring singly on long stalks at or just above the water's surface.

**Bellflower family (Campanulaceae)**
Members of this family of annual and perennial herbaceous plants usually have simple, alternate leaves and often contain milky sap. Flowers are regular or irregular, usually united at the base and tipped with five lobes. Although lobelias are considered members, some botanists place *Lobelia* species in their own family, the Lobeliaceae.

**Pink family (Caryophyllaceae)**
The leaves of this family of annual and perennial herbaceous plants are usually opposite, simple, and entire. Flowers are regular, usually with five petals (rarely four or petals absent) that may be lobed, often with five or ten stamens, one to many in branched flower clusters. Also known as the carnation family, it includes chickweed as well as carnation and baby's breath of the florist trade.

**Spiderwort family (Commelinaceae)**
These herbaceous annual and perennial plants have leaves that are simple, entire, and alternate, with a sheathing or clasping stem. Flowers are regular or irregular, with three petals, in a branched or umbel-like cluster, often with a bract just below. Most species are tropical.

**Lily-of-the-valley family (Convallariaceae)**
These perennial herbaceous plants were formerly included in the lily family (Liliaceae). Leaves are basal, alternate or whorled, simple, and entire. Flowers are regular, with four, six, or eight tepals, solitary and axillary or sometimes branched in terminal or axillary clusters. The frequently cultivated lily-of-the-valley (*Convallaria majalis*) is a European native plant.

**Morning glory family (Convolvulaceae)**

This family includes annual and perennial herbaceous plants that are typically vining or trailing. Leaves are usually simple and alternate, often lobed. Flowers are regular and usually bell-shaped, with five mostly shallow lobes, solitary in leaf axils or in clusters. The morning glory species that grow in the agricultural fields of southern Wisconsin are tropical American in origin.

**Gourd family (Cucurbitaceae)**

These annual herbaceous vines usually have coiled tendrils. Leaves are usually alternate and palmately lobed. Flowers of separate sexes are regular and usually five-lobed. Fruit is usually a berry or pepo (fleshy with a hard, inseparable rind). Common garden plants in the family include cucumber, watermelon, and squash.

**Sundew family (Droseraceae)**

This is a family of very specialized carnivorous, perennial, herbaceous plants that usually consist of a basal circle of modified leaves whose surfaces are covered with sticky, gland-tipped hairs that trap and digest small prey. Flowers are regular, five-petaled, and solitary, or several are clustered atop the flowering stem. There are about 200 species of sundews, and they grow on all continents except Antarctica.

**Heath family (Ericaceae)**

In Wisconsin, this family consists of perennial herbs and shrubs. Leaves are mostly simple and alternate or opposite. Flowers are regular, generally bell- or urn-shaped, with four or five separate or fused petals. Also called the heather family, the Ericaceae contains well-known plants such as blueberries, cranberries, and rhododendrons.

**Pea family (Fabaceae)**

This large group of plants is also called the bean or legume family because of its leguminous fruits. Most Wisconsin species are annual and perennial herbaceous plants, but some are shrubs and trees. Leaves are usually alternate and compound. Flowers are typically irregular, with five petals typically consisting of a large upper petal (the banner or standard), two lateral petals (wings), and two lower petals that are often fused (or meet at their edges) and called a keel.

**Gentian family (Gentianaceae)**

In Wisconsin, these annual and perennial herbaceous plants have simple, entire leaves that are often opposite on the stem or whorled. Flowers are regular, with four- or five-lobed petals, solitary or in clusters. Flowers of some gentian species possess some of the purest blue coloration in the plant world. The houseplant Persian violet (*Exacum affine*) is a member of the family.

**Geranium family (Geraniaceae)**

Leaves can be opposite, alternate, or basal for these mostly annual and perennial herbs. Flowers are usually regular with five petals, in clusters atop the stem. The fruit has a long beak like a crane. The family comprises hundreds of species. Most of the cultivated geraniums (*Pelargonium* species) have origins in South Africa.

**Iris family (Iridaceae)**

This is a family of perennial herbaceous plants that usually have basal, simple, sword-shaped leaves in a fanlike arrangement. Flowers are regular or nearly so, with six tepals, solitary or clustered. The commonly cultivated bearded irises of home gardens are horticultural variants involving the German iris (*Iris germanica*) of Europe.

**Mint family (Lamiaceae)**

Plants in this large family are often aromatic (though several species are not). Square stems are characteristic. Leaves are opposite and mostly simple. Flowers are usually irregular (sometimes nearly regular), tubular, and two-lipped with five lobes. The most aromatic species treated in this book are wild mint (*Mentha* species), wild bergamot (*Monarda* species), and mountain mint (*Pycnanthemum* species).

**Bladderwort family (Lentibulariaceae)**

These are carnivorous herbaceous plants of aquatic or saturated soils. Leaves of Wisconsin species are basal and flat (*Pinguicula* species) or dissected and alternate or whorled, bearing saclike bladders (*Utricularia* species). Flowers are irregular, two-lipped, with a spur at the base, and located along and atop the stem. Common butterwort (*Pinguicula vulgaris*) is restricted to areas near Lake Superior.

**Lily family (Liliaceae)**

This family of perennial herbaceous plants has a variety of leaf arrangements. Flowers are regular with six tepals, occurring solitary, paired, or in clusters. The lily family once consisted of several different genera, but recent taxonomic changes have resulted in many fewer. In this book, trout lilies (*Erythronium* species) and true lilies (*Lilium* species) are treated.

**Loosestrife family (Lythraceae)**

Most members of this family of annual and perennial herbs and shrubs grow in moist or wet habitats in Wisconsin. Leaves are mostly opposite and entire. Flowers are mostly regular, with a floral cup of four to six petals that are commonly pink and crinkled.

**Miner's lettuce family (Montiaceae)**

Plants in this family are somewhat succulent perennial herbs with simple, entire, and mostly opposite or basal leaves. Flowers are regular; the Wisconsin species are five-petaled. Spring beauty (*Claytonia* species) is placed in this book in the Montiaceae, but some authors place it in the purslane family (Portulacaceae).

**Myrsine family (Myrsinaceae)**

These annual and perennial herbs have simple, opposite or whorled leaves, some of which possess visible dots or streaks. Flowers, which may also display dots or streaks, are regular, four- to nine-parted, tubular, and funnel- or bell-shaped. In our region, most species occupy wetland habitats. A few members are currently placed by some authors in the primrose family (Primulaceae).

**Four-o'clock family (Nyctaginaceae)**

This ornamental family of perennials is characterized by mostly simple and entire opposite leaves, the pairs unequal or equal in size. Flowers are regular, and petals are absent but with five petal-like sepals, often clumped together above the sepal-like bracts. The ornamental *Bougainvillea* species of the tropics is in this family.

**Water lily family (Nymphaeaceae)**

Plants of this family are distinctive, with their large, floating or emergent leaves usually notched into two lobes. The single, commonly large, and showy flowers are regular, composed of numerous petals and sepals, the latter sometimes petal-like.

**Evening primrose family (Onagraceae)**
These biennial and perennial plants have opposite, alternate, or whorled leaves that are either entire or toothed. Flowers are often showy, normally four-parted, and regular, the floral tube elongated. Fuchsia, a mostly tropical plant often used in hanging baskets, belongs to this family.

**Orchid family (Orchidaceae)**
This family is often considered the largest plant family on Earth in terms of species numbers. They are perennial herbs, mostly tropical, with leaves (if present—some are leafless) being alternate, opposite, or whorled. Flowers have three petals, one of which (the lip) is typically larger, dissimilar in shape, and usually positioned lowermost.

**Broomrape family (Orobanchaceae)**
This large family of many genera was formerly placed in the figwort family (Scrophulariaceae). They are root parasitic (or partially so) annuals and perennials, possessing chlorophyll or not. Leaves are simple or lobed and alternate, with some reduced to scales. Flowers are irregular, two-lipped, and five-lobed.

**Wood sorrel family (Oxalidaceae)**
Plants in this family are perennial herbs with mostly palmately compound leaves having three rounded leaflets notched at the tips, basal or alternate on stem. Flowers are regular, with five petals and sepals. They are known for containing oxalic acid in their tissues. The tropical starfruit (*Averrhoa carambola*) available in some supermarkets is a member of this family.

**Poppy family (Papaveraceae)**

This family is known for having colored or milky sap in several species, such as the red sap in bloodroot (*Sanguinaria canadensis*). Plants are herbaceous, with simple and entire or compound basal or alternate stem leaves. Flowers are regular or irregular, with most having four or more petals.

**Lopseed family (Phrymaceae)**

This family includes perennial herbs with simple, opposite, entire or toothed leaves. The flowers are generally irregular, tubular, and two-lipped, with the upper lip two-lobed and the lower three-lobed. Most of the members were formerly included in the figwort family (Scrophulariaceae).

**Plantain family (Plantaginaceae)**

This family has been revised to include species previously placed in the figwort family (Scrophulariaceae). With some reservation, I adhere to that treatment. Plants are annual and perennial herbs. Leaves are basal, alternate or opposite, entire or toothed. Flowers are regular or irregular, generally four- to five-parted.

**Phlox family (Polemoniaceae)**

This well-known family consists mostly of annual and perennial herbs. Leaves are simple or compound, alternate or opposite, or basal. Flowers are regular or irregular, trumpet or bell-shaped, and five-lobed. One species naturalized in Wisconsin, garden phlox (*Phlox paniculata*), is widely cultivated.

**Milkwort family (Polygalaceae)**
These annual and perennial herbs have simple, mostly alternate and entire leaves. Flowers are generally irregular with five sepals, with the lateral two (wings) larger and petal-like; three or five true petals are small and fused to the stamen tube.

**Buckwheat family (Polygonaceae)**
This family consists of annual and perennial herbs. Leaves are generally simple, alternate, and entire. In most of our species, papery sheaths (ocreae) wrap around swollen stem nodes. Flowers are generally small and regular, with two to six tepals, often in two whorls. Garden rhubarb (*Rheum rhabarbarum*) is a member of this family.

**Pickerelweed family (Pontederiaceae)**
These perennial herbs of aquatic or wetland habitats have leaves that are simple, alternate or whorled, and linear, round, or arrowhead shaped. Flowers are regular or irregular, tubular, with six tepals.

**Buttercup family (Ranunculaceae)**
This large and highly variable family comprises annual and perennial herbs with leaves both basally positioned and alternate or opposite, simple or compound. Flowers are generally regular, with three to six sepals with some petal-like; petals are absent to several. Most if not all members contain chemicals that are toxic to humans.

**Rose family (Rosaceae)**
Members are annual, biennial, and perennial herbs, shrubs, and trees with simple or compound, mostly toothed, alternate leaves. Flowers are usually regular and five-petaled. Many well-known edible species are in the family, such as apple, blackberry, peach, and strawberry, as well as ornamentals such as crabapple, rose, and spiraea.

**Coffee family (Rubiaceae)**
Also known as the bedstraw or madder family, the plants best known in the family are coffee (*Coffea* species). Ranging from annual and perennial herbs to woody shrubs, most plants have entire, opposite leaves. Flowers are generally regular and tubular with four- or five-lobed petals.

**Sandalwood family (Santalaceae)**
Our single species of this family is an herbaceous perennial that is partially parasitic on roots of other plants. Leaves are mostly simple and entire, alternate. Flowers are regular, with five tepals.

**Pitcher plant family (Sarraceniaceae)**
This is a family of amazing carnivorous perennials that grow in wetlands. Leaves are typically modified as tubular, water-filled traps conducive to harvesting insects. Flowers are regular and nodding, with five sepals and five dangling petals surrounding a five-lobed, disk-like style.

**Saxifrage family (Saxifragaceae)**
This family of perennial herbs usually has simple leaves in a basal circle, or if leaves are on stems, then alternate, entire or toothed. Flowers are mostly regular with five petals commonly narrowed at their bases. Many members of the family grow primarily on rocks.

**Figwort family (Scrophulariaceae)**
This family includes biennial and perennial herbs. Leaves are alternate or opposite, entire or toothed. Flowers are regular or irregular, with four- or five-parted petals. Many genera formerly included in this family have been assigned elsewhere, mainly to the plantain family (Plantaginaceae).

**Trillium family (Trilliaceae)**
This is a family of perennial herbs with solitary flowers. Its large, leafy bracts (generally viewed as leaves) occur mostly in whorls of three and are simple, entire, and ovate or obovate to elliptical. True leaves are underground, alternate, and scale-like along the horizontal stem. The flower is regular with three petals. Previously, species in this family were placed in the lily family (Liliaceae).

**Cattail family (Typhaceae)**
Plants of this family are herbaceous perennials commonly growing in dense colonies in wetlands. Leaves are basal and alternate on the stem, two-ranked, and narrowly linear and strap-shaped, flat or triangular in cross section with sheathing bases. Flowers are minute and densely packed, without petals or sepals, the male and female blossoms separate on the same plant.

**Nettle family (Urticaceae)**
This family consists of annual and perennial herbs. Leaves are simple and opposite or alternate. Flowers are minute, lacking petals, with four or five sepals, with separate sexes on the same plant. Plants in this family are known for their stinging hairs, but not all species have them.

**Vervain family (Verbenaceae)**

These plants are annual and perennial herbs, many with four-angled stems. Leaves are simple, opposite, with margins entire, toothed, or deeply lobed or divided. Flowers are regular to irregular, tubular, with five lobes, often on long, narrow spikes.

**Violet family (Violaceae)**

This family consists of annuals and perennials with basal or alternate stem leaves that are simple, entire to toothed or lobed. Flowers are irregular, five-petaled, with the lowest petal often the largest and prolonged into a spur.

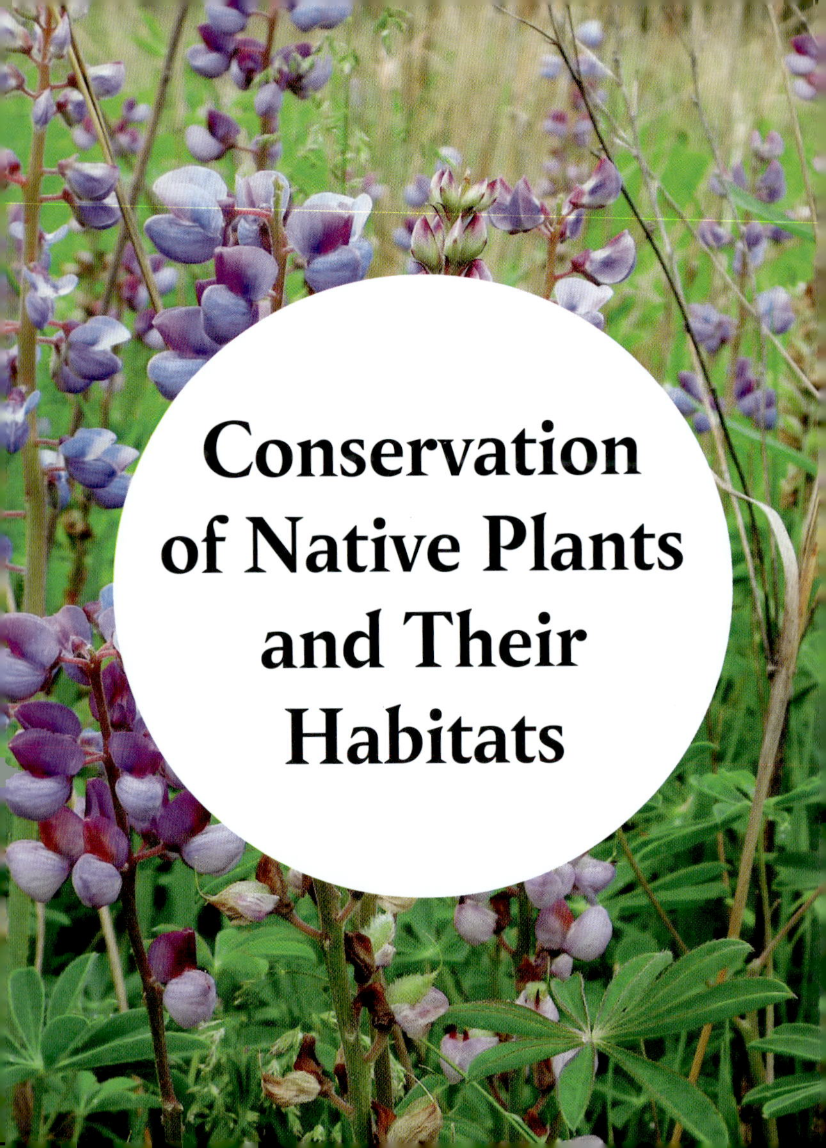
Conservation
of Native Plants
and Their
Habitats

**Though wildflowers may seem abundant** in places, their numbers today are only a fraction of what they once were. And the decline continues. Although Wisconsin has a fair amount of natural land where a good diversity of wildflowers grow, a seemingly constant barrage of impacts depletes it. One great way to help ensure that more is not lost is by supporting land protection efforts made by various public agencies and private organizations. Or if you're fortunate enough to own some natural land, you can try to keep it natural. But simply owning land does not guarantee protection. Factors such as the absence of fire (in fire-adapted natural communities) and the loss of large predators (especially those that prey on—and would thus keep in check—the state's overpopulation of deer) have caused a decline in our native flora. Equally damaging is the huge impact caused by exotic invasive plants. To minimize this impact, you can join efforts to control invasive species or avoid using them in your own landscaping.

You can also refrain from collecting wildflowers. Taking just one may seem harmless enough, but think of the impact if too many people took "just one." Wildflowers are best left to grow and be enjoyed in their native haunts. It's important to leave them in place because of the organisms that rely on them. Native insects, especially pollinators, play an important role in the web of life. Several are generalists, being involved with a whole host of plants, but some are specialists, interacting with only one plant species. If that species disappears, so does the insect. A perfect example is the association of butterflies and their host plants. Most people know about the host-specific relationship between monarchs and milkweeds—that is, monarch caterpillars feed only on the leaves of milkweeds. So if milkweeds disappear, so do these beautiful butterflies. Few people know about

the small but beautiful Karner blue butterfly and the lupines upon which its caterpillars feed. Wisconsin has the largest population of Karner blues in the world! And why is that? Because Wisconsin is home to a relatively large population of native lupines to support them.

# Using This Book

**This book introduces readers** to the wildflowers of Wisconsin, primarily using photographs. The photos are grouped by flower color and then alphabetically by family, genus, and species. Common names are also included. Pay close attention not only to the shape and color of the flowers but also to the arrangement of them on the plant. Also notice the foliage—for example, whether the leaves are opposite or alternate—as this can be as diagnostic as the flower. The following categories with specifics about the plant are included to provide further help in identification.

**HABITAT** The environment where the plant grows. Knowing the habitat of your wildflower is quite important, as some plants are restricted to a certain habitat and are found only there. For example, some plants live only in deep water or grow only on rock.

**BLOOM** The season when a wildflower will most likely be in bloom. Because there is often no precise cutoff between seasons, a range may be provided, such as "summer, fall." Be aware that blooming time may fluctuate somewhat between years, as weather is often a factor.

**DESCRIPTION** The plant's overall growth form and stature.

**FLOWER** The flower's color, overall shape and size, position, and arrangement.

**LEAF** The location and position of leaves on the plant, plus their size and characteristics of shape, margin, and surfaces.

**FRUIT** The type and distinguishing features. In some cases, seeds are mentioned.

**NARRATIVE** Provides comparisons with similar looking and/or related species, familiar species of the family in which it belongs, names and various uses of specific plants made by Native Americans and early colonists, origin of common and scientific names, trivia, and more. (Information regarding Native American plant use is derived from several published accounts, including interviews of tribal members made by Huron Smith of the Milwaukee Public Museum and others in the early 1900s.)

# White Flowers

*Sagittaria latifolia*
AlisMataceae

# common arrowhead





**HABITAT** Marshes, pond and lake margins, streams, swamps, ditches

**BLOOM** Summer

**DESCRIPTION** Erect perennial herb, hairless, often in colonies, to 3 ft. tall

**FLOWER** White, three-petaled, broad, female and male flower usually separate on same plant, to 1 in. wide, in up to ten whorls on flowering stem

**LEAF** Basal, blades arrowhead-shaped, highly variable in width of lobes, toothless, hairless, on long stalks, to 12 in. long and half as wide

**FRUIT** Flat nutlet with a short, bent beak

The six known species of arrowheads in the state can be difficult to differentiate based just on leaf shape. A better way is noting differences in the mature fruit. Many Indigenous Americans have used the tubers for food, as do wildlife, especially beaver and muskrat. Also known as duck potato, but its tubers are not a common food source for ducks.

*Allium tricoccum*
Alliaceae

# wild leek

**HABITAT** Rich, moist forests

**BLOOM** Summer

**DESCRIPTION** Erect perennial herb, single-stemmed, hairless, to 12 in. tall, commonly forms large colonies

**FLOWER** White, about ¼ in. wide, with six tepals, up to 50 in a rounded cluster to 2 in. wide

**LEAF** In early spring up to three leaves, to 12 in. long and 4 in. wide, elliptic, with distinct reddish stalk (see bottom photo) and strong onion odor; leaves no longer present when plant blooms in late spring

**FRUIT** Capsule three-chambered, each bearing a shiny black seed

Known to the Ojibwe as *bûgwa' djijîca' gowûnj* (unusual onion). Also currently known as ramps in some areas, it is becoming rare from overharvesting. A closely related but less common species is the narrow-leaved wild leek (*A. burdickii*), which blooms earlier and has fewer flowers, with a leaf base that tapers gradually and lacks reddish coloration.

*Cicuta maculata*
Apiaceae

# water hemlock





**HABITAT** Swamps, fens, shores, ditches, various wet sites

**BLOOM** Summer

**DESCRIPTION** Branched biennial/perennial herb with green, purple, or purplish blotched hollow stems, hairless, to 6 ft. tall

**FLOWER** White, five-petaled, about ⅛ in. wide, numerous in rounded, somewhat flat-topped clusters at stem tips

**LEAF** Compound, with leaflets to 4 in. long, margins sharply toothed, veins ending in notches between teeth, a trait not found in most plants

**FRUIT** Roundish, ribbed, each containing two seeds

Water hemlock is perhaps the most poisonous native plant of North America. It is reported that if one eats it, especially the roots, death can occur within a matter of minutes. It is related to poison hemlock (*Conium maculatum*), an introduced species from Europe found here mostly in disturbed areas in the southern part of the state. Poison hemlock was used to put Socrates to death.

*Cryptotaenia canadensis*
Apiaceae

# honewort





**HABITAT** Moist forests, stream terraces, ravines

**BLOOM** Summer

**DESCRIPTION** Erect, branching perennial, hairless, to 3 ft. tall

**FLOWER** White, five-petaled, to 1⁄16 in. wide, in spreading clusters at stem tips

**LEAF** Alternate, compound with three leaflets, to 6 in. long and slightly more than half as wide, irregularly toothed

**FRUIT** Narrow, ribbed, and two-seeded, seeds resemble fennel seeds

A member of the carrot family, honewort is one of the few forest plants to bloom during summer. The flowers are quite tiny. A close relative, mitsuba (*C. japonica*), is used in salads and soups in Japan. Honewort was used to treat hone, or swelling of the cheek.

*Daucus carota*
Apiaceae

# Queen Anne's lace

**HABITAT** Fields, roadsides, fencerows, disturbed sites

**BLOOM** Summer, fall

**DESCRIPTION** Erect biennial herb with some branching, hairy, to 4 ft. tall

**FLOWER** White, with five rounded petals, to ⅛ in. wide, the middle flower often dark purple, in a large, flat-topped cluster of hundreds of flowers, to 5 in. wide, with dissected leafy bracts beneath the cluster

**LEAF** Alternate, lacy, divided into multiple lance to linear segments, to 6 in. long

**FRUIT** Oblong, ribbed and bristly, about ⅛ in. long

This familiar plant of fields and roadsides is native to Eurasia. It is a member of the carrot family and, in fact, is the same species as our edible garden carrot, which was selectively bred to create the carrot of today. The root of Queen Anne's lace is comparatively small and generally lacks the color and taste of garden carrots.

*Heracleum maximum*
Apiaceae

# cow parsnip

**HABITAT** Wet meadows, thickets, floodplains, streambanks

**BLOOM** Spring, summer

**DESCRIPTION** Large, upright perennial herb, hairy, stems strongly ridged and hollow, to 8 ft. tall

**FLOWER** White, five-petaled, each with a notched tip, the outer petals larger than the inner ones, to ¼ in. wide, in flat-topped, grouped clusters, to 10 in. wide

**LEAF** Compound with three leaflets, to 15 in. long and about as wide, lobed and toothed, alternate with a papery sheath clasping the main stem

**FRUIT** Flattened, somewhat oval, about ¼ in. wide, containing a pair of seeds

Leaves and flower heads of this impressively large plant are hard to miss. One source reports it as one of the top ten plant species used for a variety of reasons by Indigenous Americans. It also contains compounds that, when in contact with skin and exposed to light, may cause severe dermatitis.

*Osmorhiza longistylis*
Apiaceae

# aniseroot

**HABITAT** Rich, moist forests

**BLOOM** Spring

**DESCRIPTION** Erect perennial herb, branched, hairless or hairy, stem commonly reddish in color, to 3 ft. tall

**FLOWER** White, five-petaled, to ⅛ in. wide, styles longer than the petals, up to 50 in. flat-topped clusters at stem tips

**LEAF** Fernlike, subdivided into leaflets that create an overall triangular outline, margins toothed, alternate, hairless or somewhat hairy, produces aroma of anise when crushed, to 9 in. long

**FRUIT** Narrow, slightly curved, tapered at both ends, to about 1 in. long, with persistent styles to ⅛ in. long

Aniseroot is recognized in part by its flat-topped cluster of flowers, but also by its roots and leaves that smell like anise. The Ojibwe have used it to make tea to treat sore throats. The similar looking sweet cicely (*O. claytonii*) differs by having little or no anise scent and shorter styles.

*Apocynum androsaemifolium*
Apocynaceae

# spreading dogbane





**HABITAT** Dry savannas, open woodlands, prairies, fields, roadsides

**BLOOM** Summer

**DESCRIPTION** Perennial herb with spreading branches, stems smooth and commonly reddish, milky sap, to 3 ft. tall

**FLOWER** White to pinkish exterior, interior with pink stripes, bell-shaped with five recurved lobes, nodding, assembled in clusters at branch tips

**LEAF** Oval, unlobed, opposite, to 4 in. long and half as wide, often drooping

**FRUIT** Resembles skinny, dangling string beans, to 6 in. long

Compared to spreading dogbane, common dogbane (*A. cannabinum*) is usually taller and has all white flowers. Their stems are a source of fiber that can be used to make cordage (rope). Its outer rind has been used as thread for fine sewing by the Ojibwe. Dogbanes are a food source for the beautiful dogbane beetle (*Chrysochus auratus*).

*Calla palustris*
Araceae

# wild calla





**HABITAT** Shallow, mostly acidic water in bogs, swamps, lake margins, slow-moving streams, ditches

**BLOOM** Spring

**DESCRIPTION** Colony-forming perennial, hairless, to 15 in. tall

**FLOWER** Greenish white, tiny, several on a thick cylindrical spike that emerges from base of a broad, white, petal-like spathe (blade), to 4 in. long and 2 in. wide

**LEAF** Basal, heart-shaped, shiny, with prominent midvein and numerous side veins, to 5 in. long

**FRUIT** Cluster of bright red (when ripe) berries

Differs from the similar looking calla lily (*Zantedeschia* species) of the florist trade. Neither species is a member of the lily family (Liliaceae), though the two are related—both occur in the arum family (Araceae) along with other Wisconsin natives such as jack-in-the-pulpit and skunk cabbage.

*Aralia nudicaulis*
Araliaceae

# wild sarsaparilla





**HABITAT** Moist forests, woodlands, swamps, dunes

**BLOOM** Spring, summer

**DESCRIPTION** Erect, three-pronged perennial, to 2 ft. tall, colony-forming

**FLOWER** White or greenish, five-petaled, to ⅛ in. wide, numerous, usually in three rounded clusters, each to 2 in. wide, atop a naked (leafless) stem

**LEAF** Pinnately compound, with normally five leaflets, these to 5 in. long and 2 in. wide, finely toothed, copper to bronze color and shiny when young

**FRUIT** Bluish black berry to ¼ in. wide

Named for its use as a substitute for sarsaparilla. Its berries are eaten by several forest birds including white-throated sparrow and ruffed grouse and are an important food for black bears during summer and fall foraging. The plant has been used medicinally for a variety of ailments and is related to ginseng (*Panax quinquefolius*), a popular medicinal herb.

*Achillea millefolium*
Asteraceae

# yarrow





**HABITAT** Fields, roadsides, grasslands, woodlands

**BLOOM** Summer

**DESCRIPTION** Erect perennial herb, usually unbranched except at top, somewhat woolly, often clumping, to 2½ ft. tall

**FLOWER** White, also pink or red, in compact heads to ½ in. wide, composed of outer ring of five white ray flowers and a central disk of up to twenty tiny tubular flowers, arranged in flat-topped clusters atop the stem

**LEAF** Finely dissected, feathery or fernlike, to 6 in. long and 1 in. wide, alternate, aromatic when bruised

**FRUIT** Small, dry one-seeded nutlet

Presumably native, but some plants may be introduced from Eurasia (where the species is also native). The genus *Achillea* is named for the Greek god Achilles, who is said to have used yarrow to treat his wounds. For North American Indigenous people, yarrow is reported to have had more medicinal uses than any other plant.

*Ageratina altissima*
Asteraceae

# white snakeroot

**HABITAT** Moist to dryish forests, open woodlands, thickets

**BLOOM** Late summer, fall

**DESCRIPTION** Erect perennial, stems mostly smooth and hairless, branching in flower clusters, to 3 ft. tall

**FLOWER** White, tiny and tubular, up to 20 in. heads about ¼ in. wide, occurring in somewhat flat-topped clusters

**LEAF** Triangular with sharply toothed margins, opposite, to 6 in. long and 4 in. wide

**FRUIT** Slender nutlet with tuft of hairs at one end that enables it to be carried in the wind

This species is attributed to having killed Nancy Hanks Lincoln, Abraham Lincoln's mother. It contains a toxic chemical that when eaten by a cow is transferred to her milk and to the person who drinks it, causing "milk sickness." It was a significant cause of death for many pioneers. *Eupatorium rugosum* is its former scientific name.

*Antennaria neglecta*
Asteraceae

# field pussytoes

**HABITAT** Prairies, fields, open woods, roadsides

**BLOOM** Spring

**DESCRIPTION** White-woolly, mat-forming perennial herb with erect flower stalks to 12 in. tall, spreads by runners

**FLOWER** White, tiny, male and female on separate plants, several in heads to ⅓ in. wide

**LEAF** Mostly basal, rather narrowly spoon-shaped with a single vein running through the middle, white-woolly beneath

**FRUIT** Seedlike fruit bears a fluffy tuft of hairs that facilitates wind dispersal

There are four species of pussytoes and several varieties in the state, which can be difficult to distinguish from one another. Its common name reflects the appearance of the furry flowering heads, somewhat like the hairy pads of a cat's foot. The genus name alludes to the male flower's club-tipped bristles, said to resemble insect antennae.

*Erigeron philadelphicus*
Asteraceae

# Philadelphia fleabane

**HABITAT** Moist forest clearings and edges, streambanks, roadsides, trails, meadows, fields

**BLOOM** Spring, summer

**DESCRIPTION** Erect, hairy biennial or perennial herb, branched above, to 3 ft. tall

**FLOWER** White to pinkish, in compact heads to ¾ in. wide, composed of an outer ring of 100 or more threadlike ray flowers and a central disk of about an equal number tiny yellow tubular flowers, in spreading clusters atop the stem

**LEAF** Basal leaves coarsely toothed, stem leaves usually less so or not at all, bases usually clasping the stem, alternate, to 3 in. long and 1 in. wide

**FRUIT** Dry, about 1⁄16 in. long, with tuft of long hairs that facilitates wind dispersal

Fleabanes reportedly are used to eradicate or repel fleas, though no evidence proves that they do either. Compared to other fleabanes in the state, this species differs by having clasping leaves. Its Ojibwe name is *mîcao'gacan* (odor of deer hooves). Smoke from its burning flowers is said to attract bucks.

*Podophyllum peltatum*
Berberidaceae

# mayapple

**HABITAT** Moist upland forests, ravines, stream terraces, thickets

**BLOOM** Spring

**DESCRIPTION** Erect perennial herb, unbranched (except for fertile plants), hairless, colonial, to 2 ft. tall

**FLOWER** White with six to nine rounded petals, occurring singly and nodding from base of stalks of paired leaves, to 1½ in. wide

**LEAF** Umbrella-like, single with stem attached to central part of blade, or double-leaved when flowering, deeply divided and tipped with large teeth, to about 12 in. wide

**FRUIT** Round berry, yellow when ripe, about 2 in. wide

Mayapple's ripe fruit mostly consists of seeds with a mucilaginous coating. They are considered edible, tasting like passionfruit to some people. Wildlife must like it, given the difficulty of finding ripe fruit. Other parts of the plant are poisonous. White slant-line moths (*Tetracis cachexiata*) use the flowers as camouflage during the day to hide from predators.

garlic mustard

*Berteroa incana*
Brassicaceae

# hoary alyssum

**HABITAT** Fields, pastures, roadsides, disturbed areas, especially in sandy soil

**BLOOM** Spring, summer, fall

**DESCRIPTION** Erect, grayish green, densely hairy annual or short-lived perennial herb, branched toward tip, to 3 ft. tall

**FLOWER** White, about ¼ in. wide, with four deeply lobed petals (appears as eight), in rounded clusters about 2 in. wide at stem tips

**LEAF** Basal (when young, absent as plant matures) and alternate on stem, hairy, unlobed, toothless

**FRUIT** Round to elliptic, somewhat flattened, with a narrow beak at tip

This European introduction begins to bloom in spring and continues into fall as its flowering stalk elongates and produces flowers. Another white-flowered invasive plant of the mustard family is garlic mustard (*Alliaria petiolata*), which is mostly a forest plant and blooms only in the spring. It has triangular-shaped leaves that smell of garlic when crushed.

*Capsella bursa-pastoris*
Brassicaceae

# shepherd's purse





**HABITAT** Disturbed areas, roadsides, vacant lots, agricultural fields

**BLOOM** Spring, summer, fall

**DESCRIPTION** Somewhat branched annual, to 2 ft. tall, with simple and star-shaped hairs toward its base

**FLOWER** White, four-petaled, to ⅛ in. wide, clustered at stem tips; one of the earliest plants to bloom

**LEAF** Basal rosette, hairy, deeply lobed, stem leaves lance- or arrowhead-shaped, alternately attached with clasping bases

**FRUIT** Triangular to sharply heart-shaped, the narrow end at base, presenting a distinctive purse-like appearance

This Eurasian native is now a worldwide weed. When wet, seeds produce a toxic, mucilaginous coating that attracts and kills various organisms such as nematodes. Compounds from the decaying nematodes are reported to benefit the plant's growth, and thus some botanists suggest shepherd's purse could be categorized as a carnivorous plant.

*Cardamine concatenata*
Brassicaceae

# cutleaf toothwort

**HABITAT** Rich, moist forests, ravines, slopes, stream terraces

**BLOOM** Spring

**DESCRIPTION** Erect perennial herb, unbranched, mostly hairless, to 12 in. tall

**FLOWER** White, four-petaled, to ½ in. wide, in loose cluster atop the stem

**LEAF** Basal, on stem usually three in a whorl, three- to five-parted, toothed, to 5 in. long and about as wide or wider

**FRUIT** Narrow, upright pods splitting lengthwise in halves

Toothwort literally means "tooth plant" for the toothlike bumps on the underground rhizome of some closely related species (not so much on this species). The rhizomes are said to have a peppery taste, somewhat like radish. Not surprisingly, both toothwort and radish are in the mustard family.

*Saponaria officinalis*
Caryophyllaceae

# bouncing-bet

**HABITAT** Fields, roadsides, railroads, savannas

**BLOOM** Summer, fall

**DESCRIPTION** Erect perennial herb, branched or not, hairless, colonial, to 3 ft. tall

**FLOWER** White to pinkish, five-petaled, broadest at tip and commonly notched, spreading or reflexed, its base in a long, straight tube formed by the sepals, to about 1 in. wide, arranged in dense clusters atop the stem

**LEAF** Oval to elliptic, stalkless, with three prominent veins, opposite, to 4 in. long and about half as wide

**FRUIT** Many-seeded capsule, splitting longitudinally from tip

This Eurasian native is also known as soapwort—*Saponaria* comes from *sapo*, meaning "soap," reflecting the ability of its juice to form a lather when mixed with water. Different explanations exist for the name bouncing-bet, including being a term used for an English washerwoman or barmaid. Double-flowered plants, used for horticultural purposes, are also known to exist in the wild.

*Silene latifolia*
Caryophyllaceae

# white campion





**HABITAT** Disturbed sites, roadsides, railroads, pastures, ditches, waste places

**BLOOM** Summer, fall

**DESCRIPTION** Erect or ascending annual or short-lived perennial, branched above with sticky, glandular-tipped hairs when mature, to 3 ft. tall

**FLOWER** White, to 1 in. wide, with five, two-lobed petals attached at tip of inflated calyx (when mature); male and female flowers on separate plants; calyx of male plants have 10 reddish, hairy veins, 20 in females, at tips of spreading branches, opening in evening

**LEAF** Lance-shaped to elliptic, unlobed, glandular hairy, opposite, to 5 in. long and 1¼ in. wide, smaller leaves commonly in axils of main leaves

**FRUIT** Tan capsule with five spreading, split teeth (looks like ten) at tip

This Eurasian species belongs to a group of plants known as catchflies. Some species have a clammy feel due to sticky glands that often trap tiny insects, leading some to consider this a primitive form of carnivory. One source says *Silene* is from *sialon* (Greek for "saliva"), referring to the plants' sticky secretions.

*Maianthemum canadense*
Convallariaceae

# Canada mayflower

**HABITAT** Dry to moist woods, even in dense shade, swamps, bogs

**BLOOM** Spring, summer

**DESCRIPTION** Erect, perennial herb with a somewhat zigzagging stem, hairy or not, to 8 in. tall

**FLOWER** White, four tepals, spreading or recurved, to ¼ in. wide, scattered on stalk atop the plant

**LEAF** Heart-shaped, alternate, clasping, two or three per plant, hairy or not, to 3 in. long and 2 in. wide

**FRUIT** Berry, when ripe, is red and speckled, about ¼ in. wide

The genus name is from the Latin *Maius* (May) and the Greek *ánthemon* (flower), giving us the common name mayflower. It is sometimes called wild lily-of-the-valley for its superficial resemblance to the garden plant *Convallaria majalis*. The Ojibwe name, *agoñgosi' mînûn*, means "chipmunk berries." One use was to cure headaches.

*Maianthemum racemosum*
Convallariaceae

# Solomon's plume





**HABITAT** Moist to dryish woods, stream terraces, thickets

**BLOOM** Spring, summer

**DESCRIPTION** Arching perennial herb with somewhat zigzagging stem, unbranched, hairless to finely hairy, to 2½ ft. tall

**FLOWER** White, with six spreading tepals and six conspicuous stamens longer than the tepals, to ¼ in. wide, in a feathery, somewhat triangular shaped cluster of flowers at the stem tip

**LEAF** Elliptic, noticeably parallel-veined, somewhat on same plane, stalkless, alternate, to 6 in. long and 3 in. wide

**FRUIT** Round berry, red when ripe

Because the foliage of Solomon's plume resembles that of smooth Solomon's seal (*Polygonatum biflorum*), another name is false Solomon's seal. Also in Wisconsin is starry Solomon's plume (*M. stellatum*), with tepals that are longer than the stamens and narrower and slightly clasping leaves.

*Calystegia sepium*
Convolvulaceae

# hedge bindweed

**HABITAT** Thickets, streambanks, fencerows, ditches, wooded edges, weedy places

**BLOOM** Summer, fall

**DESCRIPTION** Twining, hairless perennial vine without tendrils, to 10 ft. long

**FLOWER** White, pink, or bicolored, funnel-shaped, shallowly five-lobed, to 3 in. wide at mouth, with two broad green bracts at the flower base

**LEAF** Alternate, arrowhead-shaped, with angular lobes at base, hairless, to 4 in. long

**FRUIT** Rounded capsule, about ½ in. wide

This species is similar looking and related to morning glory (*Ipomoea* species), but hedge bindweed flowers differ by having two large bracts at their bases. Interesting common names include old man's nightcap, wedding gown, and belle of the ball. The specific epithet, *sepium*, means "of hedges" The flower color is quite varied. Plants are poisonous.

*Cuscuta gronovii*
Convolvulaceae

# common dodder





**HABITAT** Grows on other plants usually on wet sites—swamps, bottomland forests, fens, streambanks

**BLOOM** Summer

**DESCRIPTION** Twining annual vine, yellowish orange, branching, hairless, to 6 ft. or longer

**FLOWER** White, bell-shaped, with five spreading, slightly rounded lobes, to ⅛ in. wide, in clusters on side branches

**LEAF** Minute, scale-like if present, alternate, and same color as stem

**FRUIT** Rounded capsule, somewhat greenish and swollen at tip

These curious plants are parasites that use small suckerlike structures to withdraw water and nutrients from stems of jewelweeds, asters, and other plants. They resemble strands of cooked spaghetti draped over their hosts. Fortunately, they rarely kill the host. This one is the most widespread of at least seven species found in the state, including some that are quite rare.

peeled dried fruit

*Echinocystis lobata*
Cucurbitaceae

# wild cucumber





**HABITAT** Floodplain forests, thickets, forest edges, fencerows

**BLOOM** Summer, fall

**DESCRIPTION** Climbing annual vine with coiling tendrils, stems angled and hairless, to 10 ft. long

**FLOWER** White, somewhat cup-shaped, with six narrow, strap-shaped petals, male and female flowers separate on same plants, males several in elongated clusters, females fewer on shorter stalks, both to ½ in. wide

**LEAF** Alternate, with five triangular lobes radiating like fingers on a hand, hairless, to 4 in. long and about as wide

**FRUIT** Oblong, smooth-skinned fruit with soft prickles, to 2½ in. long

Also known as balsam apple, though its fruit is not considered edible. When thoroughly dried and peeled, fruit resembles a miniature luffa sponge. The genus name is from the Greek *echinos* (hedgehog or sea urchin) and *cystis* (bladder), a reference to its inflated spiny fruit. The Ojibwe have made tea from its roots to treat stomach troubles.

sundew flower

*Drosera rotundifolia*
Droseraceae

# round-leaved sundew





**HABITAT** Sphagnum bogs, fens, marshes, shores, wet sand prairies

**BLOOM** Summer

**DESCRIPTION** Small carnivorous herb to about 2 in. wide, essentially stemless except for flowering stalk, to 10 in. tall

**FLOWER** White, five-petaled, to ½ in. wide, blooming in ascending order along the side of an unfurling leafless stem

**LEAF** Basal rosette, often reddish, modified into round pads at ends of stalks, covered with hairs tipped with sticky droplets

**FRUIT** Egg-shaped capsule

There are four species of sundew in Wisconsin. All have sticky hairs that trap small insects that slowly dissolve to provide nutrients to the plant. This species occurs in northern latitudes around the world. Darwin wrote of it, "Is it not curious that a plant should be far more sensitive to a touch than any nerve in the human body!"

*Monotropa uniflora*
Ericaceae

# Indian pipe





**HABITAT** Moist forests, usually in acidic soils

**BLOOM** Summer, fall

**DESCRIPTION** Erect perennial herb, translucent white to pinkish, fleshy, unbranched, hairless, to 8 in. tall

**FLOWER** White, bell-shaped, to ¾ in. long, with three to six petals nodding and then turning vertical with age, occurring singly atop stem

**LEAF** Reduced to scales, alternate, white, to ⅔ in. long and ¼ in. wide

**FRUIT** When mature, an erect, black, round capsule

This amazing plant lacks chlorophyll and the ability to photosynthesize. It survives because its roots connect to fungi that supply it with food drawn from roots of green plants. Also known as ghost flower because of its pale white appearance. It is called *xawiska* (flowers white) by the Ho-Chunk people, who have used it to revive someone who has fainted.

*Pyrola elliptica*
Ericaceae

# large-leaved shinleaf





**HABITAT** Moist to dry forests, woodland slopes, in acidic soils

**BLOOM** Summer

**DESCRIPTION** Erect perennial herb, unbranched, hairless, grows low to the ground, to 12 in. tall when flowering

**FLOWER** White, about ½ in. wide, with five oval-shaped, nodding petals bearing green veins, and a downward pointing style with an upcurved tip, several positioned alternately on a flowering stalk to 12 in. tall

**LEAF** Basal, elliptic, thin, usually not very shiny, hairless, margins often with minute teeth, to 3 in. long and half as wide

**FRUIT** Dry, round capsule

Wisconsin has five species of *Pyrola*, with the large-leaved shinleaf apparently the most common. The whole group is often referred to as shinleaf. It is reported that their leaves contain an aspirin-like substance that will relieve pain when ground up and applied as a paste to shins (and, it would seem, to other body parts).

ground fruit pod

*Amphicarpaea bracteata*
Fabaceae

# American hog peanut



**HABITAT** Moist to dryish forests, thickets, seeps

**BLOOM** Summer, early fall

**DESCRIPTION** Twining herbaceous annual vine, variously hairy, to 6 ft. long; two varieties, one with purple flowers and spreading stem hairs (var. *comosa*), the other with white flowers and few hairs (var. *bracteata*), with some overlap in features between the two

**FLOWER** White or purple, typical pea family shape, to ¾ in. long, in clusters on long stalks extending from leaf axils

**LEAF** Alternate, divided into three leaflets, oval or diamond-shaped, hairy or smooth, to 4 in. long and 3 in. wide

**FRUIT** Pod, similar to a garden pea, with edible seeds; or a round pod that grows at ground level

Produces two types of flowers, one as pictured in the photo, the other lacking petals. The latter type forms an edible, round "peanut" at ground level or underground. The Ojibwe have used the seeds from the aboveground pods for food.

*Trifolium repens*
Fabaceae

# white clover

**HABITAT** Lawns, pastures, roadsides, creekbanks, gardens, disturbed places

**BLOOM** Spring, summer, fall

**DESCRIPTION** Perennial herb, grows along the ground via trailing stems that root at nodes (where leaves are attached), unbranched, mostly hairless, to 10 in. tall

**FLOWER** White, tubular, with a typical pea flower shape, five-petaled, several in a rounded head, to ¾ in. wide, positioned on a single leafless stalk

**LEAF** Compound, alternate, with three broadly elliptic leaflets, unlobed, margins finely toothed, often with a whitish chevron patch, to 1 in. long and about as wide

**FRUIT** Small pod hidden within the withered flower

The genus name of this Eurasian native reflects the three leaflets—*tri* (three) and *folium* (leaf). Many of the attributes of red clover are also present in white clover, including nitrogen fixation and use as a forage crop. White clover is the most likely clover to be growing in lawns.

*Mentha canadensis*
Lamiaceae

# wild mint

**HABITAT** Marshes, swamps, fens, meadows, shores, ditches

**BLOOM** Summer, fall

**DESCRIPTION** Erect to sprawling perennial herb, square-stemmed, usually unbranched, variously hairy (sometimes mostly on the angles), with a strong minty scent when bruised, rhizomatous, to 2 ft. tall

**FLOWER** White to pale purple, tubular, with two flaring lips, the upper one with two lobes, the lower one with three, to ⅛ in. long, densely packed in whorls around the leaf axils

**LEAF** Elliptic with sharp teeth, opposite, pairs at 90 degrees to one another, to 2½ in. long and ¾ in. wide

**FRUIT** Small, dry nutlets

Also known as *M. arvensis*, this is Wisconsin's only native *Mentha* mint. Several other *Mentha* species are also found here, introduced mostly from Europe. Water horehound (*Lycopus* species) looks similar but lacks a minty scent. Wild mint has often been used by Native Americans for medicinal purposes. It is a natural source of menthol.

*Pycnanthemum virginianum*
Lamiaceae

# common mountain mint





**HABITAT** Prairies, marshes, fens, meadows, fields

**BLOOM** Summer, fall

**DESCRIPTION** Erect perennial herb, branched on upper plant, four-angled with hairs on the angles, often colonial, to 3 ft. tall

**FLOWER** White, to ⅛ in. wide, tubular, two-lipped, purple-spotted, upper lip shallowly notched, lower lip three-lobed, in flat heads to ¾ in. wide

**LEAF** Narrowly lance-shaped, tapering to a point, hairless, unlobed, toothless, and stalkless, opposite, very aromatic when crushed, to 2½ in. long and ½ in. wide

**FRUIT** Dry, capsule-like with four nutlets

*Pycnanthemum* is from the Greek *pyknos*, meaning "dense," and *anthemon* for "flower," referring to its densely packed flower heads. Insects, especially butterflies, are strongly attracted to mountain mint flowers. It is sometimes used as an ornamental, perhaps because it is said to be deer proof.

*Erythronium albidum*
Liliaceae

# white trout lily

**HABITAT** Moist woods, floodplains, slopes, ravines

**BLOOM** Spring

**DESCRIPTION** Erect to reclining perennial herb, unbranched, hairless, colonial by stolons, to 7 in. tall

**FLOWER** White, but may be pinkish tinged, bell-shaped, with six reflexed tepals occurring singly atop the flowering stem

**LEAF** Elliptic to lance-shaped, somewhat pliable and waxy looking, pale or mint-green, usually mottled with brown or purplish splotches; flowering plants have two leaves, nonflowering (the majority) only one

**FRUIT** Oval capsule, usually held erect

This species is scattered throughout the state. Dwarf trout lily (*E. propullans*), also white flowered, occurs in far southeastern Minnesota and nowhere else on Earth. Given that state's proximity to Wisconsin, it might occur here as well. Its flowers are considerably smaller than those of white trout lily and often have only four or five tepals.

*Claytonia virginica*
Montiaceae

# Virginia spring beauty

**HABITAT** Moist to dryish forests

**BLOOM** Spring

**DESCRIPTION** Erect to ascending perennial herb, unbranched, hairless, to 6 in. tall (usually much shorter)

**FLOWER** White, about ½ in. wide, five-petaled, commonly with pink stripes, loosely arranged atop an elongating flowering stalk

**LEAF** Basal, plus a single pair oppositely arranged on flowering stem, linear to narrowly elliptic, somewhat fleshy, stalks indistinct, to 4 in. long and ½ in. wide

**FRUIT** Round capsule cradled in a cuplike pair of persistent sepals

Spring beauty belongs to a class of plants called spring ephemerals. They arise from the ground in early spring before the leaves of trees appear. Their tops die back soon after the tree leaves fully develop. Carolina spring beauty (*C. caroliniana*) occurs in the far northern counties. Each leaf has a distinct stalk.

*Nymphaea odorata*
Nymphaeaceae

# fragrant white waterlily

**HABITAT** Ponds, lakes, calm water portions of rivers, often rooted in highly organic soil

**BLOOM** Summer, fall

**DESCRIPTION** Aquatic perennial herb with branching rhizomes from which leaves grow, the stalks of the latter to 6 ft. long

**FLOWER** White, solitary on long stalks, many petaled, floating to perched slightly above the water surface, fragrant, to 7 in. wide

**LEAF** Mostly round with a slit on one side from which most main veins spread palmately, shiny, often purple underneath, to 12 in. long and wide

**FRUIT** Berry-like, fleshy, matures underwater on coiled stalks

The name *Nymphaea* comes from the mythological water nymph, a female deity that lives in watery places, the same environment where waterlilies occur. With its many lakes, Wisconsin is the perfect place for waterlilies. (Smaller leaves in picture are watershield.) Flowers open in the morning and close in the evening.

*Circaea canadensis*
Onagraceae

# enchanter's nightshade

**HABITAT** Moist forests, ravines, stream terraces, thickets

**BLOOM** Summer

**DESCRIPTION** Relatively small, erect perennial herb, sparsely hairy on at least upper stem and branches, to 2 ft. tall

**FLOWER** White, approximately ⅛ in. wide, with two notched petals, giving the impression of four petals

**LEAF** Opposite pairs, these at right angles to adjacent leaf pairs, lance-shaped, mostly hairless, with scattered teeth along margin, to 4 in. long and about half as wide

**FRUIT** Capsule, grooved on the surface, bearing hooked hairs that enable it to attach to clothing and fur

Fruits of this plant stick tight to clothing or fur via tiny hooks that grab onto passersby. The similar alpine enchanter's nightshade (*C. alpina*) is more delicate and lacks grooves on the fruit. These two species are not true nightshades, or at least they are not members of the nightshade family (Solanaceae).

squirrel corn

*Dicentra cucullaria*
Papaveraceae

# Dutchman's breeches





**HABITAT** Rich, moist forests, wooded ravines, stream terraces

**BLOOM** Spring

**DESCRIPTION** Erect perennial herb, unbranched and hairless, with pinkish white underground bulblets, to 12 in. tall

**FLOWER** White, to ¾ in. long, with two inconspicuous petals and two large, inflated, upwardly spreading petals that resemble pointed spurs (the "legs"), dangling upside down on an arching stem, like miniature trousers on a clothesline

**LEAF** Basal, highly dissected into narrow lobes, giving a feathery appearance, hairless, grayish green, pale beneath

**FRUIT** Narrow capsules that taper at each end, resembling tiny pea pods

Flowers of Dutchman's breeches are similar in appearance to the less common squirrel corn (*D. canadensis*, see inset), but the latter's two larger petals are not spreading and its underground bulblets are yellow. The Asian bleeding-heart (*D. spectabilis*) is an ornamental relative.

*Sanguinaria canadensis*
Papaveraceae

# bloodroot





**HABITAT** Rich, moist, forested ravines, stream terraces, slopes

**BLOOM** Spring

**DESCRIPTION** Erect perennial herb, unbranched, hairless, from a rhizome with blood-red sap, to 10 in. tall

**FLOWER** White, with a ring of up to 12 elliptic petals surrounding numerous yellow stamens, solitary, to 2 in. wide

**LEAF** Single, rounded in outline with up to seven deep lobes at the tip, rubbery in texture, hairless, to 6 in. long and about as wide

**FRUIT** Capsule; seeds with gelatinous food body attached

Named for its blood-red sap, *Sanguinaria* is from the Latin *sanguis* (blood) and *sanguinarius* (bleeding). Each seed has a food body attached that attracts ants. The ants take the seeds to their nests, consume the gelatinous food bodies, and discard the seeds, leaving them to germinate and potentially form new plants.

*Phryma leptostachya*
Phrymaceae

# American lopseed



**HABITAT** Moist to dryish forests, ravines, stream terraces

**BLOOM** Summer

**DESCRIPTION** Erect perennial herb, mostly unbranched, hairy, four-angled stems, to 2 ft. tall

**FLOWER** White, tubular, about ¼ in. long, two-lipped, the small upper lip pinkish purple, notched in the middle and curved upward, the lower lip three-lobed, occurring in pairs evenly spaced on stalks atop plant and in leaf axils, initially held horizontally, drooping with age

**LEAF** Oval, toothed, opposite, pairs positioned at 90 degrees to one another

**FRUIT** Dry nutlet with hooks at the tip, pressed down against the stem

Flowers are often overlooked because of their small size. Its distinctive fruit is tightly appressed downward (like ears of a lop-eared rabbit), thus the common name. Each fruit has three hooks at one end, suggesting a mechanism for attaching to fur or clothing, but they rarely seem to do so.

*Chelone glabra*
Plantaginaceae

# white turtlehead

**HABITAT** Marshes, fens, stream borders, seeps, forested swamps

**BLOOM** Late summer, fall

**DESCRIPTION** Erect perennial herb, hairless, with square stem, to 4 ft. tall

**FLOWER** White, some with pink-tinged tips, two-lipped, tubular, several in cluster at stem tip

**LEAF** Opposite, narrow, spear-shaped with toothed margins, to 6 in. long and 1½ in. wide

**FRUIT** Round to oval capsule, brown when mature

Both common and Latin names reference the flower shape, like that of a turtle's head. *Chelone* is Greek for "tortoise." Although typically all white, some flowers may have a blush of pink at the tips. White turtlehead occurs in every county of the state.

*Penstemon digitalis*
Plantaginaceae

# foxglove beardtongue





**HABITAT** Prairies, fields, roadsides, open woodlands, clearings

**BLOOM** Summer

**DESCRIPTION** Erect perennial herb, unbranched, mostly hairless, to 3 ft. tall

**FLOWER** White, tubular, to 1 in. long, the tube narrow then abruptly inflated toward the opening; five-lobed, two upper and three lower, usually covered with glandular hairs; flowers of some plants with purple streaking within, in a cluster of upright branches atop the plant

**LEAF** Basal leaves stalked and oblong, to 7 in. long and 2 in. wide; those on stem opposite, clasping, lance-shaped, broadest at base, unlobed, mostly toothless, hairless

**FRUIT** Conical capsule bearing numerous seeds

Beardtongues are named for the hairy, or "bearded," sterile anther located within the throat of the flower. There are nearly 300 species of beardtongue species worldwide, of which almost 240 occur in North America, most in the western United States. Beardtongues are popular in wildflower seed mixes used for native plant restoration.

*Veronicastrum virginicum*
Plantaginaceae

# Culver's root

**HABITAT** Prairies, savannas, open woodlands, meadows, fields

**BLOOM** Summer

**DESCRIPTION** Erect perennial herb, unbranched, mostly hairless, to 6 ft. tall

**FLOWER** White, tubular, to ⅓ in. long, with four short lobes and long exerted pollen-bearing stamens, numerous in vertical, candelabra-like spikes to 8 in. long

**LEAF** Lance-shaped in whorls of four or five, toothed, to 5 in. long and 1 in. wide

**FRUIT** Elliptical to egg-shaped capsules with numerous seeds

Native Americans and early colonists used the plant to treat tuberculosis. Referred to as "Culver's root" in a letter dated 1716 by Puritan Cotton Mather to John Winthrop, it is likely named for an acquaintance of Mather's identified in the letter as "Mr. Culver" from Lebanon, Connecticut. Why it's named for him is unclear.

*Actaea rubra*
Ranunculaceae

# red baneberry

**HABITAT** Moist upland forests and woodlands, forested swamps

**BLOOM** Spring

**DESCRIPTION** Erect, branched perennial, to 3 ft. tall

**FLOWER** White, about ¼ in. wide, four to eight petals, short-lived, persistent stamens white and showy, numerous in a rounded cluster to 4 in. long, at tip of stalk well above foliage

**LEAF** Compound, two or three times divided, leaflets about 2 in. long and wide, coarsely toothed, undersurface somewhat hairy on veins

**FRUIT** Several seeded, shiny red (rarely white) berries on slender green stalk

Poisonous to humans, especially its berries, but the latter are eaten by birds and small mammals with no apparent harm. The similar white baneberry, or dolls eyes (*A. pachypoda*), has berries that are typically white and rarely red, and the stalk of each is usually much thicker than that of red baneberry.

*Anemone canadensis*
Ranunculaceae

# Canada anemone





**HABITAT** Moist sedge meadows, wet prairies, shorelines, floodplains, road banks

**BLOOM** Spring, summer

**DESCRIPTION** Short-hairy, rhizomatous perennial herb, to 2 ft. tall, colonial

**FLOWER** White, to 1½ in. wide, five sepals are petal-like (true petals absent)

**LEAF** Basal and leaflike bracts deeply three- to five-lobed and toothed, basal leaves on long stalks, bracts opposite and stalkless, all to 6 in. across

**FRUIT** Flattened seed with a noticeable beak, several packed into a tight cluster

This attractive wildflower, also known as meadow anemone, has tempted many to consider using it in their home landscaping, but it can aggressively spread. The Ojibwe name is *mîdewidji' bîk* (medicine lodge root), as its roots have been used as throat lozenges to help one sing better in the medicine lodge ceremony.

false rue anemone

rue anemon

*Anemone quinquefolia*
Ranunculaceae

# wood anemone

**HABITAT** Moist to relatively dry forests, swamps, thickets, streambanks

**BLOOM** Spring

**DESCRIPTION** Erect, low-growing perennial herb, hairy, unbranched, colonial by rhizomes, to 8 in. tall

**FLOWER** White, usually with five elliptic, petal-like sepals, to 1 in. wide, single on a slender stalk positioned above the leaves

**LEAF** Usually in a single whorl of three, each with mostly three lobes, but the outer two segments can be split, giving the overall impression of five pointed segments per leaf, hairy, to 2 in. long and about half as wide

**FRUIT** Oval seed with a noticeable beak, several packed into a rounded cluster

The epithet means five-leaved (but see leaf description). Other white-flowering forest plants in Wisconsin that bloom at about the same time are false rue anemone (*Enemion biternatum*), with five petal-like sepals, and rue anemone (*Thalictrum thalictroides*), which usually has more. Both have hairless leaflets with somewhat rounded bases and tips.

*Clematis virginiana*
Ranunculaceae

# virgin's bower





**HABITAT** Forest edges, thickets, fencerows, streambanks

**BLOOM** Summer

**DESCRIPTION** Climbing perennial vine to 15 ft. long, branching with sparsely hairy and angled stems

**FLOWER** White, with four petal-like sepals, about ¾ in. wide, male and female flowers on separate plants

**LEAF** Compound, in opposite pairs on stem, three leaflets, sparsely hairy and coarsely toothed, to 3 in. long and about as wide

**FRUIT** Seedlike with an attached feathery plume, clustered in a rounded, fuzzy ball

Virgin's bower is related to the commonly cultivated ornamental clematis. Two native species are in Wisconsin, this one and purple clematis (*C. occidentalis*). The latter is less common and mostly restricted to the northern counties and Driftless Area. As its common name implies, its flowers are purple, and are also larger and drooping.

*Hepatica americana*
Ranunculaceae

# round-lobed hepatica





**HABITAT** Dry to moist forests

**BLOOM** Spring

**DESCRIPTION** Erect perennial herb, unbranched, hairy (especially when young), to 6 in. tall

**FLOWER** White, but often purplish pink, to 1 in. wide, with 5–11 petal-like sepals in a ring around a center of numerous stamens, true petals absent, solitary on hairy stalks, several per plant

**LEAF** Basal, several per plant, three-lobed, each rounded at the tip, thick textured, mottled on the surface, purplish underneath, persists through winter

**FRUIT** Flattened, elliptic nutlet

Some consider this species and sharp-lobed hepatica (*H. acutiloba*) (both also known as liverleaf) as varieties of *H. nobilis*. Both occur in the state. The main differences between the two species are the shape of the leaf lobes and habitat preference. Round-lobed hepatica tends to prefer drier sites. Both were used to treat liver ailments. *Hepatica* is from *hepaticus*, "of the liver."

*Fragaria virginiana*
Rosaceae

# wild strawberry

**HABITAT** Prairies, savannas, open woodlands, forest edges, fields, roadsides

**BLOOM** Spring

**DESCRIPTION** Low-growing perennial herb producing runners that root at the tips to form colonies of new plants, to 6 in. tall

**FLOWER** White, to 1 in. wide, five-petals encircle a cluster of yellow stamens and pistils on a receptacle that will develop into a strawberry, several in a group at tip of flowering stem

**LEAF** Compound with three leaflets, mostly oval-shaped, sharply toothed, to 3 in. long and half as wide

**FRUIT** Fruiting receptacle (strawberry) fleshy, round, and red, with numerous dark, seedlike fruits each sunken in a pit

Fruits are small and sweet, much sweeter than garden strawberries. Another Wisconsin native, the woodland strawberry (*F. vesca*), is cone-shaped like the garden strawberry, with the "seeds" positioned on the surface (not in pits). The Ojibwe named it *ode' imîn* (heart berry), a reference to its shape.

*Galium boreale*
Rubiaceae

# northern bedstraw



**HABITAT** Dry to wet sites, forests, various wetlands, thickets, ditches

**BLOOM** Summer

**DESCRIPTION** Erect perennial herb, branched with four-sided stem, usually smooth and hairless, occasionally hairy, to 3 ft. tall

**FLOWER** White, short, tubular, with four-pointed, spreading lobes, each on stalks in clusters atop plant and upper leaf axils, to ¼ in. wide

**LEAF** Narrow, slightly widest near base, hairy along margins, to 2 in. long and ¼ in. wide, bearing three prominent veins, in groups of four in evenly spaced whorls

**FRUIT** Ball-shaped, two-lobed (appears as two separate balls), hairy or not

Northern bedstraw is one of the showier bedstraws. It is in the madder plant family, as is coffee. Perhaps not surprisingly, the roasted seeds of a Wisconsin relative, cleavers (*G. aparine*), have been considered by some to be the best substitute for coffee of our northern plants.

berry

*Mitchella repens*
Rubiaceae

# partridgeberry

**HABITAT** Acid soils of dry to moist forests, usually in places with minimal leaf litter

**BLOOM** Summer

**DESCRIPTION** Creeping perennial subshrub, branching, growing along ground and forming mats, evergreen, to 12 in. long

**FLOWER** White, trumpet-shaped, with four spreading or recurved lobes with long hairs on inner surface, occurring in pairs, to ½ in. long

**LEAF** Round, glossy, with whitish midvein, evergreen, opposite, to ¾ in. long and about as wide

**FRUIT** Red, berry-like, formed by fusion of ripened ovaries of the paired flowers (see inset)

Partridgeberry often grows on rocks, mossy hummocks, steep slopes, and other areas with minimal leaf litter and competition from other plants. Although its fruit looks inviting, it's dry and tasteless. The pair of scars on the berry are remnants of the two flowers involved in its formation.

*Comandra umbellata*
Santalaceae

# bastard toadflax





**HABITAT** Dry prairies, savannas, dunes, open woodlands, rocky shores

**BLOOM** Spring, summer

**DESCRIPTION** Erect, colonial perennial herb, hairless, partially parasitic, to 18 in. tall

**FLOWER** White, bell-shaped with five petal-like sepals, to ¼ in. long

**LEAF** Elliptic, hairless, untoothed, to 2 in. long and ½ in. wide, alternate

**FRUIT** Single-seeded with a thin, fleshy outer coat, reported to be somewhat edible

Although it produces some of its food through photosynthesis, bastard toadflax also parasitizes several different plant species through a connection with their roots. It is one of the few (only?) plant species that occurs coast to coast, linked by adjoining counties across the entire United States. Interestingly, it also occurs in the Balkan Peninsula. It is an alternate host for a blister rust that affects pine trees.

flower

*Mitella diphylla*
Saxifragaceae

# bishop's cap





**HABITAT** Moist forests, ravines, streambanks

**BLOOM** Spring

**DESCRIPTION** Erect perennial herb, unbranched, hairy, flower stalk to 12 in. tall

**FLOWER** White, with five deeply fringed petals, to ¼ in. wide, several on slender upright stalk

**LEAF** Basal, with a single pair on the flowering stem; basal leaves with up to five sharp lobes, to 4 in. long and about as wide, stem leaves three-lobed, opposite at midstem, to 2½ in. long

**FRUIT** Capsules, splitting to expose a collection of shiny black seeds

Flowers resemble snowflakes, but one of the species' common names (miterwort), as well as its scientific one, refer to its fruit shape, which is reminiscent of a miter, a cap worn by bishops. The seeds are spread when rain splashes them out of the cap. A close relative, naked miterwort (*M. nuda*), has greenish flowers and a leafless flowering stalk. It occurs mostly in the northern counties.

nodding trillium

*Trillium grandiflorum*
Trilliaceae

# large-flowered trillium





**HABITAT** Moist forests, stream terraces, ravines, slopes

**BLOOM** Spring

**DESCRIPTION** Erect perennial herb, unbranched, hairless, to 20 in. tall

**FLOWER** White, somewhat funnel-shaped, three petals, broadly triangular and slightly bent at tip, overlapping at their bases, turning pink with age, to 3 in. wide, solitary on an upright stalk

**LEAF** Leaves (technically leaflike bracts) broad, sharp-tipped, hairless, to 6 in. long and about as wide, in a whorl of three atop the stem

**FRUIT** Fleshy, berry-like seeds, dispersed by ants

Of the six *Trillium* species in the state, this one is by far the largest and showiest. Somewhat similar is nodding trillium (*T. cernuum*, see inset). Its flower is smaller and commonly bends out of sight below the leaves. White-tailed deer eat trillium plants and can significantly reduce their numbers.

# Pink to Red Flowers

*Asclepias incarnata*
Apocynaceae

# swamp milkweed

**HABITAT** Marshes, wet meadows, shorelines, ditches

**BLOOM** Summer

**DESCRIPTION** Erect perennial herb with milky sap, hairless, to 4 ft. tall

**FLOWER** Pink, about ¼ in. wide, consisting of five reflexed petals, pinkish white hoods, and incurved horns, numerous in somewhat flat-topped clusters to 3 in. wide

**LEAF** Narrowly lance-shaped, entire, hairless, to 6 in. long and 1 in. wide, opposite

**FRUIT** Narrow upright pods with smooth surfaces, to 4 in. long; seeds within have silky tufts of hairs attached that assist in wind dispersal

Of the several species of milkweed found in Wisconsin, swamp milkweed is the one most likely to be found in wet sites. It is also one of the most common species. The Latin genus name is from Asclepius, Greek god of medicine and healing. The epithet *incarnata* references the flesh color of the flowers.

*Asclepias syriaca*
Apocynaceae

# common milkweed

**HABITAT** Prairies, fields, roadsides, savannas, ditch banks

**BLOOM** Summer

**DESCRIPTION** Upright perennial herb, unbranched, short-hairy, with milky sap and long rhizomes, to 6 ft. tall

**FLOWER** Pale pink, consisting of five reflexed petals, pinkish white hoods, and incurved horns, to ½ in. wide, numerous in somewhat rounded clusters to 4 in. wide, fragrant

**LEAF** Broad, elliptic, densely hairy below, to 12 in. long and 4 in. wide, opposite

**FRUIT** Spreading pods to 4 in. long, surface hairy with soft prickles; seeds within have silky tufts of hairs that assist in wind dispersal

This is this most common milkweed in the state. It provides necessary food for monarch caterpillars and is critical to the species' survival. The fluffy seeds were once used as a filler for life jackets. Prairie milkweed (*A. sullivantii*) is similar but is hairless and restricted to the southern counties.

*Cirsium arvense*
Asteraceae

# Canada thistle, creeping thistle

**HABITAT** Fields, roadsides, ditches, fencerows, rights-of-way

**BLOOM** Summer

**DESCRIPTION** Erect perennial herb, colonial, to 4 ft. tall

**FLOWER** Pinkish purple, disk flowers only, in relatively small, crowded heads, each about ¾ in. wide, male and female flowers mostly separate on different plants

**LEAF** Lobed, mostly with sharp spines at tips, hairiness variable, alternate, to 6 in. long and 2 in. wide

**FRUIT** Seedlike, each bearing a long tuft of hairs at one end

Canada thistle is not native to Canada or North America. It's of Eurasian origin but has spread around the world as one of the most noxious weeds. It is certainly one of Wisconsin's most invasive thistles. It forms large colonies by spreading roots as well as by being a prolific seed producer. Eradicate where possible.

*Cirsium muticum*
**Asteraceae**

# swamp thistle

**HABITAT** Seeps, fens, swamps, wet meadows

**BLOOM** Summer, fall

**DESCRIPTION** Tall biennial herb, stem ridged, mostly spineless with sparse, cobwebby hairs, to 8 ft. tall

**FLOWER** Pinkish purple, disk flowers only, numerous in widely spreading heads to 1½ in. wide, the little bracts on head below flowers mostly spineless, with cobwebby hairs and often sticky to the touch

**LEAF** Deeply lobed, with relatively soft spines, alternate, to about 12 in. long and 4 in. wide

**FRUIT** Seedlike, each bearing a long tuft of hairs at one end

Swamp thistle is native and lacks the highly invasive nature of most of the nonnative thistles. It is also not viciously prickly like most thistles. There are at least six native thistles in the state, all important food sources for wildlife. They provide nectar for pollinators and seeds for birds, especially finches.

*Eutrochium maculatum*
Asteraceae

# spotted Joe-Pye-weed

**HABITAT** Marshes, bogs, fens, seep springs, wet prairies, ditches

**BLOOM** Summer

**DESCRIPTION** Erect, unbranched (below flower branches) perennial herb, hairy or nearly hairless, to 8 ft. tall, colonial

**FLOWER** Pinkish purple, disk flowers only, tubular, five-lobed, to 20 per compact head, the heads about ¼ in. wide, in a somewhat flat-topped branched cluster atop the main stem, no ray petals

**LEAF** Lance-shaped, narrowed at both ends, sharply toothed, veins deeply impressed, to 9 in. long, grouped in whorls of three to six on a vertical stem that's commonly purple-spotted

**FRUIT** Slender, one-seeded, with tuft of long hairs at one end that helps it float in the wind

Many stories exist about the origin of the name Joe-Pye weed. It was most likely named after Joseph Shauquethqueat, or "Joe Pye," a Mohican leader from the late 18th century. He lived in western Massachusetts, where the plant's name was likely first used. Why the plant is named for him is unclear.

*Liatris aspera*
Asteraceae

# rough blazing star

**HABITAT** Dry prairies, savannas, open woodlands, roadsides

**BLOOM** Summer, fall

**DESCRIPTION** Erect perennial herb, unbranched, hairy, to 4 ft. tall

**FLOWER** Pink, disk flowers only, up to 40 small, tubular five-lobed flowers in several scattered heads, to 1 in. wide, on a single spike, with tiny bulging bracts on head below flowers bending outward, their whitish to clear jagged tips curling inward

**LEAF** Narrowly lance-shaped, unlobed and toothless, rough to touch, alternate, the lowest ones to 12 in. long and 1 in. wide, becoming progressively smaller upward

**FRUIT** Small, dry, wedge-shaped nutlet with silky hairs that facilitate wind dispersal

Six species of blazing stars occur naturally in the state. All have showy pink flowers that are highly desirous to a host of butterflies, bees, and other insects. The plants typically grow in areas exposed to full sun. They are becoming popular in landscaping, especially for gardeners who want to feature native plants.

*Brasenia schreberi*
Cabombaceae

# watershield

**HABITAT** Acidic water in lakes, ponds, swamps, slow-moving streams

**BLOOM** Summer

**DESCRIPTION** Aquatic, colonial perennial rooted in soil of still bodies of water

**FLOWER** Pinkish red to purple, about ¾ in. wide, with three recurved petals and three similar sepals, on a stalk held just above the water during the day, withdrawing under the surface each evening

**LEAF** Floating on water surface, oval, somewhat football-shaped, alternate, to 5 in. long and 3 in. wide, undersurface purple; leaf undersurface and stalk have a slimy, jellylike coating

**FRUIT** Club-shaped, one- or two-seeded, in clusters

The mucilaginous coating that covers most of the plant helps protect it from feeding insects. In some Asian countries, it is eaten as a vegetable. During the first day the flower opens, it is functionally male; on the second day, it is functionally female. In some populations, this order may be reversed.

*Lobelia cardinalis*
Campanulaceae

# cardinal flower

**HABITAT** Wet forests, swamps, marshes, streambanks, ditches

**BLOOM** Summer, fall

**DESCRIPTION** Erect perennial herb, unbranched, hairless or slightly so, with milky sap, to 4 ft. tall

**FLOWER** Red, tubular, to 1½ in. long, five-lobed, three of which are positioned lowermost and two spreading laterally, numerous in a tall stalk atop the stem

**LEAF** Lance-shaped with shallow teeth, unlobed, alternate on short stalks or stalkless, to 6 in. long and 1½ in. wide

**FRUIT** Round, ribbed capsule with numerous seeds

Few plants have flowers as brilliantly red as cardinal flower. Ruby-throated hummingbirds are among its principal pollinators. It is used as a garden ornamental but requires considerable soil moisture to thrive. The common name and epithet *cardinalis* are thought to have been inspired by the red caps of Roman Catholic cardinals.

*Trifolium pratense*
Fabaceae

# red clover

**HABITAT** Pastures, fields, roadsides, ditches, disturbed areas

**BLOOM** Summer, fall

**DESCRIPTION** Erect or ascending perennial, branched, hairy, to 2 ft. tall

**FLOWER** Pinkish red to purplish, to 1½ in. wide, tubular, five-petaled, with typical pea flower shape, several in a rounded head positioned immediately above a pair of small leaves

**LEAF** Compound, alternate, with three broadly elliptic leaflets, unlobed, margins smooth or finely toothed, often with a light green chevron patch, to 2½ in. long and 1 in. wide

**FRUIT** Small pod hidden within the withered flower

This introduced Eurasian clover is a forage crop for livestock. Because it also fixes nitrogen, it is often planted as a cover crop for soil health. Regarding health, one study on female lab rats showed that red clover isoflavones may help reduce bone loss. Several other *Trifolium* clovers in Wisconsin are also nonnative and weedy.

*Monarda fistulosa*
Lamiaceae

# wild bergamot

**HABITAT** Prairies, savannas, fields, open woodlands, thickets, roadsides, usually in rather dry conditions

**BLOOM** Summer

**DESCRIPTION** Erect perennial herb, aromatic, branched, finely hairy, with square stems to 4 ft. tall

**FLOWER** Pinkish, tubular, two-lipped, the upper one narrow and unlobed, hairy, the lower one three-lobed, to 1½ in. long, several ringing a dome-like head at stem tips

**LEAF** Lance-shaped with toothed margins, on short stalks, opposite, usually hairy, to 5 in. long and 2 in. wide

**FRUIT** Dry, with usually four nutlets

Wild bergamot has a purported similarity of scent with a type of sour orange grown near the town of Bergamo, Italy. Also called bee balm for being attractive to bees, it is very minty and thus used in potpourri and sachets. Historically it was used to treat a variety of medical conditions. The Ojibwe have used it to cure catarrh and bronchial issues.

germander

*Stachys pilosa*
Lamiaceae

# marsh hedgenettle

**HABITAT** Marshes, wet prairies, meadows, stream borders, shores, ditches

**BLOOM** Summer

**DESCRIPTION** Erect perennial herb, unbranched, square-stemmed, with spreading hairs on angles as well as sides, to 3 ft. tall

**FLOWER** Pink with dark purple spots, to ½ in. long, tubular, two-lipped, upper lip hoodlike, lower three-lobed with the lowest one the longest, in several whorls on upper stem.

**LEAF** Lance-shaped, deeply veined, margins with short, sharp teeth, opposite, lacks pleasant minty aroma, to 4 in. long and about 2 in. wide

**FRUIT** Dry, capsule-like, separating into four nutlets

Sometimes known as *S. palustris*, but technically that species is European and not known in the state. The flower of germander (*Teucrium canadense*, see inset) is similar, but the upper hood is replaced by two pointed lobes. *Stachys* is Greek for "spike," probably in reference to the plant's arrangement of flowers. Interestingly, a person named Stachys is mentioned in Romans 16 of the Bible.

*Mirabilis nyctaginea*
Nyctaginaceae

# wild four-o'clock

**HABITAT** Moist and dry prairies, roadsides, fields, railroads, disturbed sites

**BLOOM** Summer

**DESCRIPTION** Erect perennial herb, branched, stem predominantly hairless and angled, to 3 ft. tall

**FLOWER** Pink, to ½ in. wide, petals absent, but with five petal-like sepals notched at the tips, in clusters at stem tips

**LEAF** Triangular with a somewhat heart-shaped base, opposite, to 4 in. long and 3 in. wide

**FRUIT** Nutlets, hairy and ribbed

So named for the opening of its flowers in late afternoon, wild four-o'clock is related to the garden version (*M. jalapa*) from South America. Wild four-o'clock can be invasive, colonizing sites far from its natural range in the Great Plains. Caterpillars of the micromoth *Neoheliodines nyctaginella* feed only on wild four-o'clocks and related species.

*Calopogon tuberosus*
Orchidaceae

# tuberous grass pink

**HABITAT** Bog, fens, swamps, wet prairies

**BLOOM** Summer

**DESCRIPTION** Erect perennial herb, unbranched, hairless, 1–2 ft. tall

**FLOWER** Pink, to 2 in. wide, three-petaled, the lip petal is uppermost with yellow club-shaped bristles, sepals pink and petal-like, in a loosely flowered cluster atop the stem

**LEAF** Typically one, basal, linear and grasslike, unlobed, hairless, to 9 in. long and 1 in. wide

**FRUIT** Erect capsule with thousands of tiny, dustlike seeds

This orchid is one of around 50 species of native orchids occurring in Wisconsin. *Calopogon* means "beautiful beard," referring to the yellow club-shaped bristles on the lip petal. The flower is unusual in the orchid family because the lip is situated at the top of the flower, rather than at the bottom. It is the only orchid in the state with that trait.

*Cypripedium acaule*
Orchidaceae

# pink lady's slipper

**HABITAT** Dry or wet acidic forests, bogs, swamps, sphagnum moss hummocks

**BLOOM** Spring

**DESCRIPTION** Erect perennial herb, unbranched, hairy, to 18 in. tall

**FLOWER** Pink, three-petaled, one shaped like a pouch with a center cleft and reddish veins, to 2½ in. long, the other two purplish, lance-shaped with a slight twist, to 1½ in. long, solitary on a single stalk emerging from the center between the basal leaves

**LEAF** Basal, in a pair, elliptic, unlobed, hairy, ribbed, to 8 in. long and 5 in. wide

**FRUIT** Ascending capsule with thousands of tiny, dustlike seeds

There are five species of lady's slippers in the state, and this is the most common, especially northward. The genus name is from the Greek *Kypris*, from Cypris, the goddess of love and beauty, and the Latin *pedis* for "foot." It is also known as moccasin flower for the shape of the large slipper-shaped lower petal. The Menominee have used it to treat male disorders.

*Cypripedium reginae*
Orchidaceae

# showy lady's slipper





**HABITAT** Fens, seepage swamps, sedge meadows

**BLOOM** Summer

**DESCRIPTION** Erect, unbranched perennial, hairy, to 3 ft. tall, often growing in clumps

**FLOWER** Pink and white, one to three flowers atop stem, three-petaled, the pinkish pouch to 2½ in. long, the other two lateral petals white, untwisted, to 2½ in. long, three sepals, white, appearing as two (two are fused)

**LEAF** Alternate, oval, noticeably ribbed, covered with short glandular hairs, to 10 in. long and 5 in. wide

**FRUIT** Erect capsule with thousands of tiny, dustlike seeds

Showy lady's slipper is Wisconsin's largest native orchid. It is indeed an impressive plant to behold, and consequently some people have transplanted it from the wild into their home gardens. This practice is often unsuccessful and such activity negatively impacts wild populations. It is best to leave them in place for all to enjoy.

*Phlox pilosa*
Polemoniaceae

# prairie phlox

**HABITAT** Moist and dry prairies, savannas, barrens, open woodlands

**BLOOM** Spring, summer

**DESCRIPTION** Erect perennial herb, multi-stemmed, mostly unbranched except above, hairy, to 2 ft. tall

**FLOWER** Pink, to ¾ in. wide, with five flaring lobes at tip of ⅔ in. long tube, the base of each often with darker pink markings, in a rounded cluster atop the plant

**LEAF** Narrowly lance-shaped, unlobed, toothless, usually hairy, opposite, the pairs positioned at 90 degrees to one another

**FRUIT** Rounded capsule with seeds thrown some distance from plant upon drying

*Phlox* is Greek for "flame," perhaps referring to the plant's twisted flower bud, shaped like a tiny torch. As natural prairie disappears in the state, so does this plant. In turn, animals that depend on it are also affected, such as the rare phlox moth (*Schinia indiana*), whose caterpillars feed only on flowers of prairie phlox.

*Polygala paucifolia*
Polygalaceae

# fringed polygala

**HABITAT** Dryish to moist forests

**BLOOM** Spring

**DESCRIPTION** Erect to trailing perennial herb, unbranched, hairless, colonial, to 6 in. tall

**FLOWER** Pinkish, to ¾ in. wide, three-petaled forming a tube, lower one fringed at the tip, and two enlarged petal-like sepals resembling wings, up to four occurring in upper leaf axils

**LEAF** Scale-like, alternate, except for larger egg-shaped ones in a whorl atop the stem with flowers, to 1½ in. long and ¾ in. wide

**FRUIT** Rounded capsule

Flowers of this interesting little plant brighten the forest floor in spring. Most are pink but some are white. In addition to the showy flowers are budlike blossoms that never open, occurring underground. They nonetheless are fertile and produce viable seed. This species is also known botanically as *Polygaloides paucifolia*. It mostly occurs in the northern counties.

*Persicaria amphibia*
Polygonaceae

# water smartweed

**HABITAT** Lakes, ponds, slow-moving streams and rivers, marshes, ditches

**BLOOM** Summer, fall

**DESCRIPTION** Erect to sprawling aquatic perennial herb, branched or not, hairy or hairless, to 4 ft. tall

**FLOWER** Pink, five tepals, to ⅛ in. wide and ¼ in. long, packed into either egg-shaped clusters (var. *stipulacea*, pictured, has floating leaves) or longer cylindrical clusters (var. *emersa* grows on exposed soil)

**LEAF** Floating on water's surface, or not, lance-shaped or elliptic, alternate and with a papery sheath where the leaf stalk attaches to a knotlike swelling on the stem, hairy or not

**FRUIT** Dry nutlet, two- or three-angled, often hidden in dried flowers

When eaten, some species of smartweeds have a peppery taste that "smarts" the tongue. Also known as knotweeds because of the swollen knots on the stem where the leaves attach. Other family members include buckwheat and rhubarb. Seeds of water smartweed are desirable food for wildlife, particularly waterfowl.

*Aquilegia canadensis*
Ranunculaceae

# wild columbine

**HABITAT** Rock outcrops, steep wooded slopes, streambanks, terraces

**BLOOM** Spring, summer

**DESCRIPTION** Multi-branched perennial with hairless stems, to 30 in. tall

**FLOWER** Red and yellow, about 2 in. long, shape unlike anything else in state, nods from arching stem tips, bears long vertical spurs with club-shaped tips, five yellow petals, five red petal-like sepals, numerous yellow stamens exerted from throat

**LEAF** Basal plus alternately attached stem leaves, doubly compound, three leaflets, each with three deeply lobed subleaflets about 2 in. long and wide

**FRUIT** Pods have long tails (beak), within which are shiny black seeds

Wild columbine is the only naturally occurring columbine in eastern North America. It likely occurs in every county of the state. The Latin *columbinus* means "dove-like," and the name reflects the allusion of doves portrayed in the inverted flower. The Meskwaki call the seeds *mänwäha* (love perfume) and have used it to empower persuasion.

# Orange Flowers

*Asclepias tuberosa*
Apocynaceae

# butterfly milkweed

**HABITAT** Prairies, old fields, savannas, roadsides, especially sandy sites

**BLOOM** Summer, fall

**DESCRIPTION** Spreading, coarsely hairy perennial herb to 3 ft. tall, with clear sap

**FLOWER** Orange, red, or yellow, about ½ in. wide, consisting of five reflexed petals, hoods, and incurved horns, numerous in somewhat rounded clusters to 3 in. wide

**LEAF** Narrow, lance-shaped, hairy, mostly alternate on stem, to 4 in. long and 1 in. wide

**FRUIT** Upright, lance-shaped pods, finely hairy, to 6 in. long; seeds within have silky tufts of hairs that assist in wind dispersal

The brilliantly colored flowers of butterfly milkweed have attracted considerable interest for use as a landscaping plant, especially in native plant gardens. It prefers well-drained soil. The plant lacks milky sap, an unusual trait for a milkweed. Like other milkweeds, it is a food host for monarch caterpillars.

yellow touch-me-not

*Impatiens capensis*
**Balsaminaceae**

# jewelweed

**HABITAT** Seeps, swamps

**BLOOM** Summer, fall

**DESCRIPTION** Branching, erect annual herb, hollow-stemmed, hairless, to 4 ft. tall

**FLOWER** Orange, cone-shaped like a cornucopia, to 1 in. long, five-lobed, open in front and narrowing to a small spur in back, spotted, dangles from upper leaf axils

**LEAF** Alternate, elliptic with wavy or bluntly toothed margins, to 5 in. long and 3 in. wide

**FRUIT** Green cylindrical capsule

A leaf submerged in water shows why this plant is named jewelweed, as it takes on a silvery sheen. Also known as touch-me-not because of the explosive nature of the ripe seed capsule when disturbed. *Impatiens* is the Latin word for "impatient," alluding to the seed, which "can't wait" to leave the capsule. Yellow touch-me-not (*I. pallida,* see inset) is similar except with yellow flowers.

wood lily

*Lilium michiganense*
Liliaceae

# Michigan lily



**HABITAT** Wet prairies, savannas, meadows, wet woods, fens, thickets

**BLOOM** Summer

**DESCRIPTION** Erect perennial herb, unbranched, hairless, to 6 ft. tall

**FLOWER** Orange, to 3 in. wide, with six reflexed, tepals speckled with dark purplish dots, nodding, up to eight on stalks atop plant

**LEAF** Narrowly lance-shaped, unlobed, untoothed, hairless, deeply parallel veined, to 4 in. long and 1 in. wide, in several spaced whorls of three to seven leaves, uppermost ones can be opposite or alternate

**FRUIT** Erect, three-angled capsule

Two lilies are native to Wisconsin, with Michigan lily being the most common. The other, wood lily (*L. philadelphicum*, see inset), is much shorter with flowers that face upward. The nonnative tiger lily (*L. lancifolium*) is similar to Michigan lily, but its leaves are alternate and the upper ones commonly have a bulblet in each axil.

*Castilleja coccinea*
Orobanchaceae

# Indian paintbrush





**HABITAT** Moist prairies (especially sandy), savannas, meadows, sandy and gravely shorelines

**BLOOM** Spring, summer

**DESCRIPTION** Erect biennial herb, partially parasitic, hairy, to 2 ft. tall

**FLOWER** Greenish yellow, tubular, to ¾ in. long, mostly obscured by large, variously lobed red, orange, or yellow bracts

**LEAF** Three-lobed, stalkless, alternate, hairy, to 3 in. long

**FRUIT** Smooth, oblong capsule

The modified leaves (bracts) below and surrounding the flowers are the colorful parts of this plant. Widespread in the state, it is one of only a few species of paintbrush in the eastern United States (many occur in the West). Indian paintbrush is partially parasitic, obtaining some but not all its nutrition from other plants through a connection with their roots.

# Yellow Flowers

*Acorus americanus*
Acoraceae

# American sweetflag





**HABITAT** Marshes, pond and lake margins, seep springs, wet prairies

**BLOOM** Spring, summer

**DESCRIPTION** Erect, unbranched, hairless perennial, to 6 ft. tall

**FLOWER** Yellowish green, very tiny, hundreds tightly packed on a fingerlike spike, to 3 in. long and ¾ in. wide, arising from the lower half of leaf

**LEAF** Slender, bright green, and sword-shaped, resembling an iris or cattail, to 4 ft. tall and ¾ in. wide, with multiple veins parallel to the larger midvein

**FRUIT** Small berry with one to six seeds

When bruised, the foliage of sweetflag is pleasantly aromatic. It was once highly prized for its medicinal properties. A similar looking sweetflag from Europe (*A. calamus*) has been introduced into North America and is apparently widespread in the state. The native species has fertile flowers, while the introduced one does not.

*Zizia aurea*
Apiaceae

# golden alexanders

**HABITAT** Moist prairies, sedge meadows, dry and moist woodlands, thickets, creek banks

**BLOOM** Spring, summer

**DESCRIPTION** Erect perennial herb, somewhat branched, hairless, colonial, to 3 ft. tall

**FLOWER** Yellow, five-petaled, to ⅛ in. wide, all on short stalks except for a central stalkless one, up to 25 in somewhat flat-topped, small clusters that together make up a larger flattish cluster to 3 in. wide

**LEAF** Compound, basal leaves twice three-parted, stem leaves sometimes less parted, leaflets lance-shaped, sharply toothed, overall outline to 6 in. long and about as wide

**FRUIT** Oval, flattened, ribbed, with two seeds

Caterpillars of the black swallowtail (*Papilio polyxenes*) feed on foliage of this species, and golden alexanders mining bees (*Andrena ziziae*) feed almost exclusively on its pollen. Most or all basal leaves in Wisconsin's other species of *Zizia*, heart-leaved golden alexanders (*Z. aptera*), are heart-shaped.

*Bidens cernua*
Asteraceae

# nodding bur-marigold





**HABITAT** Mudflats, shorelines, swamps, fens, ditches

**BLOOM** Summer, fall

**DESCRIPTION** Erect, branching, mostly hairless annual herb, to 3 ft. tall

**FLOWER** Yellow, in composite heads to 2 in. wide, outer ring of usually eight ray flowers, or absent, flowers of center disk numerous, tiny, and tubular

**LEAF** Opposite, hairless, coarsely toothed, with bases fused to the stem, to 5 in. long and 1 in. wide

**FRUIT** Flattened and brown, with four barbed awns that can stick to fur and clothing (and skin!)

Part of a group known as beggar-ticks, alluding to the fruits sticking to clothing and fur, as if begging a ride. The epithet name *cernua* means "nodding," a reference to the flowering heads that nod with age. Some plants bloom when only 1 or 2 in. tall.

*Helenium autumnale*
Asteraceae

# common sneezeweed

**HABITAT** Wet meadows, fields, fens, swamps, shorelines

**BLOOM** Summer, fall

**DESCRIPTION** Erect perennial herb, branched, variously hairy, with thin blades (wings) of tissue extending down the stem from leaf bases, to 5 ft. tall

**FLOWER** Yellow, in composite heads arranged at branch tips, to 2 in. wide, outer ring of up to 20 ray flowers with notched tips, spreading or drooping, each to ¾ in. long; central disk rounded, yellow, with hundreds of tiny, tubular flowers

**LEAF** Lance-shaped, alternate, tapered at both ends, with small, widely scattered teeth on the margins, to 5 in. long and ¾ in. wide

**FRUIT** Narrow, wedge-shaped, ribbed, about 1⁄16 in. long

Sneezeweed can cause sneezing, but not from its pollen, which is too heavy to be windborne and thus would not be inhaled. Its common name reflects the traditional use of its powdered leaves to make snuff, which brings on sneezing, and the health benefits some believe can be derived from it.

*Helianthus giganteus*
Asteraceae

# giant sunflower

**HABITAT** Marshes, wet prairies, fens, swamps, wet forests

**BLOOM** Summer, fall

**DESCRIPTION** Erect perennial herb, stem purple or green, usually with spreading hairs, to 10 ft. tall

**FLOWER** Yellow, in composite heads arranged on tips of spreading branches, to 3 in. wide, outer ring of up to 20 showy, sterile ray flowers, each to about 1¼ in. long, central disk yellow with dozens of tiny tubular flowers

**LEAF** Lance-shaped, mostly alternate on upper stem, opposite on lower stem, tapered at both ends, usually with coarse teeth on the margins, surface rough, to 6 in. long and 1½ in. wide

**FRUIT** Small, dry, one-seeded nutlet, smaller than that of garden sunflower

*Helianthus* means "sun flower," from the combination of the Greek *helios* (sun) and *anthos* (flower). There are several species in the state. Smooth oxeye (*Heliopsis helianthoides*) is a common sunflower-like plant, but each of its outer ray flowers are fertile, producing a seed (sterile in sunflower).

*Matricaria discoidea*
Asteraceae

# pineapple-weed

**HABITAT** Disturbed sites, gravel roads, cracks in pavement, parking lots, compacted soils

**BLOOM** Spring to fall

**DESCRIPTION** Erect annual herb, branched, mostly hairless, to 12 in. tall (but usually much shorter)

**FLOWER** Yellowish green, tubular, abundant in cone-shaped heads, ray flowers absent, to ⅓ in. wide, on upright stalks from upper leaf axils

**LEAF** Feathery, deeply divided in several linear segments, alternate, to 2 in. long and ¾ in. wide

**FRUIT** Tiny, dry, cylindrical nutlets

Introduced from the western states, it mostly avoids natural areas, usually confining itself instead to highly disturbed sites. The plant has a pleasant pineapple scent detected either by walking on it or crushing it with your fingers. Various Indigenous peoples have used the plant for treating gastrointestinal maladies or as a perfume or an insecticide.

*Packera paupercula*
Asteraceae

# balsam ragwort





**HABITAT** Moist prairies, meadows, fens, shores

**BLOOM** Spring, summer

**DESCRIPTION** Erect perennial herb, unbranched, mostly hairless, but may have cobwebby hairs in leaf axils and lower stem, to 1½ ft. tall

**FLOWER** In compact heads to ¾ in. wide, composed of an outer ring of up to 13 yellow ray flowers and a central disk of several tiny yellow and tubular flowers, arranged in spreading clusters atop the stem

**LEAF** Mostly basal, oval to lance-shaped with small, rounded to sharp teeth along margins, to 3 in. long and ¾ in. wide, stem leaves small, scattered, and deeply lobed

**FRUIT** Tiny, dry, with tuft of long, silky hairs at one end

There are several species of ragwort in Wisconsin, including the similar but uncommon prairie ragwort (*P. plattensis*). It is usually quite hairy and grows in dry prairie habitats. The common name appears to refer to the plant's leaves, especially the "ragged" stem leaves. The term *wort* means "plant," from the Old English *wyrt*.

*Rudbeckia hirta*
Asteraceae

# black-eyed Susan





**HABITAT** Prairies, savannas, fields, meadows, roadsides

**BLOOM** Summer, fall

**DESCRIPTION** Erect annual or biennial herb, mostly unbranched except possibly for upper plant, coarsely hairy, to 2½ ft. tall

**FLOWER** In compact heads to 3 in. wide, composed of an outer ring of up to 20 yellow ray flowers and a central cone-shaped disk of several tiny, blackish brown tubular flowers, the heads atop plant or a few from stalks in leaf axils

**LEAF** Basal leaves stalked, stem leaves usually not, lance-shaped, toothless or with few teeth, hairy, alternate, to 7 in. long and 2 in. wide

**FRUIT** Narrow, dark, wedge-shaped, four-angled nutlet

This wildflower is also a popular garden plant. The common name refers to the dark "eye" of the flowering head and may originate from an early 18th century English poem titled *Sweet William's Farewell to Black-Ey'd Susan*. Perhaps the dark disk of the flowering head reminded the colonists of the poem's lady.

*Solidago altissima*
Asteraceae

# tall goldenrod

**HABITAT** Prairies, savannas, fields, roadsides, meadows, thickets

**BLOOM** Summer, fall

**DESCRIPTION** Erect perennial herb, unbranched except for flowering branches, hairy, rhizomatous, to 6 ft. tall

**FLOWER** Yellow, to ⅓ in. wide, comprising an outer ring of up to 15 ray flowers and a central disk of three to eight tiny, tubular flowers, in a spreading, pyramidal arrangement atop the stem

**LEAF** Elliptic to lance-shaped with a few small teeth, three veined, upper surface commonly rough, alternate, to 6 in. long and 1 in. wide

**FRUIT** Brown, narrowly wedge-shaped nutlet (seed within)

One of the more common of the several species of goldenrods in the state. They get a bad rap for causing hay fever, but their pollen is too heavy to be airborne and inhaled. The most likely cause of hay fever is ragweed (*Ambrosia* species). Its pollen is dustlike and wind dispersed at the time goldenrods are in bloom.

*Taraxacum officinale*
Asteraceae

# common dandelion

**HABITAT** Lawns, fields, pastures, roadsides, waste areas

**BLOOM** Mostly spring, through fall

**DESCRIPTION** Erect perennial herb, unbranched, hairless or sparsely so, milky sap, to 15 in. tall

**FLOWER** Yellow, to 2 in. wide, usually with up to 100 ray flowers, disk flowers absent, crowded in a composite head, solitary at tip of a vertical hollow stem

**LEAF** All basal, elliptic or widest toward tip, with deeply triangular lobes and irregular teeth, to 10 in. long and 3 in. wide

**FRUIT** Small, one-seeded, brownish nutlet attached to a plume of bristles, together with others form a globe-shaped cluster (puffball) to about 1 in. wide

Dandelions were introduced from Eurasia but occur on all continents except Antarctica. The word "dandelion" appears to be from the French *dent de lion* (tooth of the lion), a reference to its toothed leaves, but in France today it's called *pissenlit* (and historically *pissabed* in England), names referring to the diuretic effects from eating the plant, leading to increased urination.

*Tragopogon dubius*
Asteraceae

# goat's beard





**HABITAT** Fields, roadsides, pastures, waste areas, disturbed areas

**BLOOM** Spring, summer

**DESCRIPTION** Erect biennial herb, gray-green, cobwebby hairs when young, hairless at maturity, milky sap, to 3 ft. tall

**FLOWER** Pale yellow, to 2 in. wide, with 100 or more ray flowers, disk flowers absent, small leafy bracts of head longer than flowers, head solitary at tip of vertical hollow stem, swollen at tip

**LEAF** Mostly basal and alternate, linear, grasslike, unlobed, mostly hairless, clasping at base, to 12 in. long and ¾ in. wide

**FRUIT** One-seeded, brownish nutlet attached to a plume of bristles, clusters in a large globe shape resembling a giant dandelion seed head, to 3 in. wide

This European native was named for its grayish, fluffy seed head, from the Greek *tragos* (goat) and *pogon* (beard). Flowers usually open in early morning and close around noon on sunny days, inspiring another common name, Jack-go-to-bed-at-noon. A cultivated (and edible) relative is salsify (*T. porrifolius*).

berry-like seeds

*Caulophyllum thalictroides*
Berberidaceae

# blue cohosh

**HABITAT** Moist, rich woods, ravines, shaded slopes

**BLOOM** Spring

**DESCRIPTION** Erect perennial herb, hairless, to 2 ft. tall, stem green or purplish, often with whitish coating that rubs off

**FLOWER** Yellow or greenish yellow, some purple-tinged, to ½ in. wide, six petal-like sepals

**LEAF** Compound, two per plant if flowering or fruiting, one if not, leaflets several, to 3 in. long with pointed lobes

**FRUIT** Fleshy, berry-like seeds, green at first, then dark blue when ripe

The Latin name *thalictroides* means "like *Thalictrum*," the genus of meadow rue. Leaves of these two unrelated plants have a similar appearance. The berry-like seeds may entice you to eat them, but they are reported to be poisonous. The Ojibwe have used the root for menstrual cramps.

*Lithospermum canescens*
Boraginaceae

# hoary puccoon

| | |
|---|---|
| HABITAT | Dry to moist prairies, savannas, open woodlands |
| BLOOM | Spring, summer |
| DESCRIPTION | Erect, perennial herb, soft-hairy, growing in clumps, to 1½ ft. tall |
| FLOWER | Yellow, tubular, with five spreading lobes, to ½ in. wide, several in a flattish cluster atop the plant |
| LEAF | Narrow, unlobed, soft-hairy with prominent midvein, alternate, to 2 in. long and ½ in. wide |
| FRUIT | Hard, white nutlet |

The genus name refers to its hard nutlets, from the Greek *lithos* (stone) and *sperma* (seed), and the specific epithet *canescens* (hoary) refers to its soft, grayish white foliage. The similar hairy puccoon (*L. caroliniense*) is rough-hairy. *Puccoon* is a common name used by some Indigenous people for plants used for dyes. Its shiny nutlets have been used as sacred beads by the Menominee.

*Barbarea vulgaris*
Brassicaceae

# yellow rocket

**HABITAT** Roadsides, fields, riverbanks, waste areas, thickets

**BLOOM** Spring

**DESCRIPTION** Erect biennial herb, branched above, mostly hairless, stem somewhat angled, to 2½ ft. tall

**FLOWER** Yellow, four-petaled, to ⅓ in. wide, in compact clusters at branch tips

**LEAF** Basal and lower leaves deeply lobed and stalked, becoming progressively smaller, unlobed and stalkless upward, alternate, the larger to 8 in. long

**FRUIT** Narrow, upright pods splitting lengthwise into halves

This early blooming wildflower of the mustard family is introduced from Eurasia. The genus name honors Saint Barbara, patron saint of people involved with explosives, such as artillerymen and miners. The plant is said to have been useful in covering and healing wounds. The Latin epithet *vulgaris* means "common." Also called wintercress, as the basal leaves overwinter.

*Clintonia borealis*
Convallariaceae

# bluebead lily

**HABITAT** Moist forests, swamps, bogs

**BLOOM** Spring

**DESCRIPTION** Erect perennial herb, unbranched, to 15 in. tall, colonial

**FLOWER** Yellow, bell-shaped, with six slender and flaring petal-like segments, to ¾ in. wide, in a cluster of up to six on a bare stem

**LEAF** Basal, glossy, somewhat thick and fleshy, oblong, with fine hairs along the margins, to 8 in. long and 3 in. wide

**FRUIT** Round, deep blue berry, about ¼ in. wide

Contrary to its common name, bluebead lily is not a true lily. Its berries are an attractive porcelain blue and although they may look appetizing, they are said to be bitter tasting and mildly toxic. The Ho Chunk call it *sumaka* (dog medicine), a treatment used to kill worms in dogs.

*Uvularia grandiflora*
Convallariaceae

# large-flowered bellwort





**HABITAT** Rich, moist forests, stream terraces, ravines, slopes

**BLOOM** Spring

**DESCRIPTION** Erect perennial herb, branched, arching, mostly hairless, often clump-forming, to 2 ft. tall

**FLOWER** Yellow, bell-shaped, to 2 in. long, six tepals, slightly twisted, drooping from stem tips

**LEAF** Elliptic, perfoliate (stem within blade tissue, as if pierced), unlobed, undersurface often hairy, alternate, to 4 in. long and half as wide

**FRUIT** Capsule, three-lobed with rounded edges

*Uvularia* is named after the uvula, that dangling tissue in the back of a person's throat. Its "pierced" leaves are distinctive. Leaves of the other bellwort in the state, sessile-leaved bellwort (*U. sessilifolia*), are not pierced. The root of large-flowered bellwort, known to the Ojibwe as *wabûckadji' bîk* (white root), has been used to treat stomach ailments.

*Lotus corniculatus*
Fabaceae

# bird's-foot trefoil

**HABITAT** Roadsides, pastures, fields, disturbed sites

**BLOOM** Summer

**DESCRIPTION** Sprawling perennial herb, branched, variously hairy or hairless, angled, to 2 ft. tall

**FLOWER** Yellow, to ½ in. long, five-petaled with a typical pea family shape, the broad uppermost banner petal may be marked with red at the base, occurring in clusters on long stalks

**LEAF** Pinnately compound with five leaflets, the lowest two separated on the stalk from the crowded upper three, oval, to ¾ in. long and half as wide

**FRUIT** Narrow, peapod-like, dark in color, to 1 in. long

Introduced from Eurasia as a cover crop and for forage, this invasive plant is a road-runner, often found along manicured highway verges. Bird's-foot trefoil is a member of the pea family and, despite its scientific name, is not closely related to American lotus (*Nelumbo lutea*), an aquatic plant with large, round leaves and whitish flowers.

*Melilotus officinalis*
Fabaceae

# yellow sweet clover

**HABITAT** Fields, roadsides, waste areas, prairies in dryish, calcareous soils

**BLOOM** Summer

**DESCRIPTION** Erect, usually biennial, branches ascending, hairless, to 5 ft. tall

**FLOWER** Yellow, five-petaled with typical pea family shape, to ¼ in. long, in loose clusters at tips of upright stalks

**LEAF** Compound, with three oval, finely tooted leaflets to 1½ in. long and ⅔ in. wide, alternate

**FRUIT** Strongly wrinkled singled-seeded pod

Despite its common name, yellow sweet clover is not a species that most people recognize as clover. That name is usually reserved for the group that includes white and red clovers. Also known as melilot, it was introduced from Europe with good intentions but has become an invasive weed. White sweet clover (*M. albus*) is similar but has white flowers.

leaves
and bladders

*Utricularia vulgaris*
**Lentibulariaceae**

# common bladderwort

**HABITAT** Lakes, ponds, swamps, slow-moving streams and rivers, ditches

**BLOOM** Summer

**DESCRIPTION** Floating aquatic perennial, carnivorous, stems to 3 ft. long

**FLOWER** Yellow, looks somewhat like a snapdragon, two-lipped, upper lip upright, the lower with a hump and a forward-projecting spur from beneath, to ¾ in. long, several on a stalk extending above the water's surface

**LEAF** Threadlike segments forking several times, with small bladders (traps) that catch and digest tiny aquatic organisms

**FRUIT** Roundish capsule holds numerous tiny seeds

This carnivorous plant has tiny bladders with hinged trap doors that, when triggered, rapidly suck in unsuspecting organisms such as water fleas and mosquito larvae. The bladders contain secretions that break down prey so that nutrients can be absorbed. Some microorganisms, however, can live in the bladder unaffected. *Utricularia macrorhiza* is another scientific name for this species.

*Erythronium americanum*
Liliaceae

# yellow trout lily

**HABITAT** Moist woods, floodplains, slopes, ravines

**BLOOM** Spring

**DESCRIPTION** Erect to reclining perennial herb, unbranched, hairless, colonial by stolons, to 7 in. tall

**FLOWER** Yellow, bell-shaped, about 1½ in. long, with six reflexed tepals, occurring singly atop the flowering stem

**LEAF** Elliptic to lance-shaped, somewhat pliable and waxy looking, pale or minty green, usually mottled with brown or purplish splotches; flowering plants have two leaves, most plants with only one leaf and nonflowering

**FRUIT** Oval capsule, usually held erect

Like other spring ephemerals, trout lilies often occur in large colonies and are present only for a few weeks in early spring. Most trout lily plants do not produce flowers. The leaves of yellow and white trout lilies look virtually alike. The trout lily mining bee (*Andrena erythronii*) is a frequent visitor to this plant's flowers.

starflower

*Lysimachia thyrsiflora*
Myrsinaceae

# tufted loosestrife

**HABITAT** Swamps, bogs, fens, floodplains, lakeshores, ditches

**BLOOM** Summer

**DESCRIPTION** Erect perennial herb, unbranched, finely hairy on upper stem, to 2½ ft. tall

**FLOWER** Yellow, to ⅓ in. wide, with five to seven spreading narrow petals and a similar number of long-exerted yellow stamens, several in oval clusters on stalks from leaf axils at middle of plant

**LEAF** Slender, lance-shaped to linear, opposite, to 4 in. long and ½ in. wide, commonly dotted on the surfaces

**FRUIT** Round, dry capsules

Tufted loosestrife occurs worldwide at far northern latitudes. Several members of the genus *Lysimachia* provide floral oil instead of nectar as food for visiting bees. All the native loosestrife species in Wisconsin have yellow flowers except starflower (*L. borealis*, or *Trientalis borealis*, see inset), which has white flowers.

*Nuphar variegata*
Nymphaeaceae

# yellow pond lily





**HABITAT** Ponds, lakes, calm water portions of rivers, often rooted in highly organic soil

**BLOOM** Summer

**DESCRIPTION** Aquatic perennial herb with thick, spongy rhizomes from which leaves grow, leaf stalks to 6 ft. long

**FLOWER** Yellow with maroon base, petals tiny and inconspicuous, usually six showy, rounded sepals, together forming what looks like a yellow cup or bowl, solitary on vertical stalks held well above the water level, to 2 in. wide

**LEAF** Shiny, oval to narrowly heart-shaped with cleft at base, usually floating level with surface (sometimes above during low water), stalks flattened on one side, to 10 in. long and 6 in. wide

**FRUIT** Berry-like, somewhat fleshy, with numerous seeds, on straight stalks

This plant, along with water lily (*Nymphaea odorata*), are often called lily pads. The leaf stalks of yellow pond lily are attached to a thick, ropy looking rhizome. The Ojibwe call it *odîte'abûg* (flat heart leaf).

*Oenothera biennis*
Onagraceae

# common evening primrose

**HABITAT** Prairies, old fields, fallow farm fields, roadsides, disturbed sites, usually on rather dry soils

**BLOOM** Summer, fall

**DESCRIPTION** Erect biennial herb, unbranched or branched above, smooth or hairy, to 6 ft. tall

**FLOWER** Yellow, to 2 in. wide, comprising four rounded petals with notched tips and four long, pointed, reflexed greenish sepals, borne on leafy spikes atop stem

**LEAF** Lance-shaped, hairy, margins toothed or not, wavy, alternate, to 7 in. long and 2 in. wide, leaves all basal first year

**FRUIT** Narrow, cylindrical, dry capsule

This rather weedy native is by far the most common of the dozen or so evening primrose species in the state. Closed for most of the day, its flowers open quickly in the evening, sometimes in less than a minute. The colorful primrose moth (*Schinia florida*) hides inside the flowers during the day, and its caterpillars feed exclusively on them.

*Cypripedium parviflorum*
Orchidaceae

# yellow lady's slipper





**HABITAT** Variety of upland forest sites in rather dry to wet soils, fens, swamps

**BLOOM** Spring, summer

**DESCRIPTION** Erect unbranched perennial, hairy, to 2 ft. tall, often growing in clumps

**FLOWER** Yellow, three-petaled, one shaped like a pouch to 2½ in. long, the other two in lateral spiraled petals to 3 in. long

**LEAF** Alternate, lance-shaped to oval, noticeably ribbed, hairy, to 9 in. long and 5 in. wide

**FRUIT** Erect capsule with thousands of tiny, dustlike seeds

Wisconsin has two varieties of yellow lady's slippers—var. *pubescens* (pictured) and var. *makasin*. The latter's flower usually has a smaller pouch and darker colored lateral petals. It is also found in wetter sites. The Ojibwe name is *ma' kasîn* (moccasin). The root is said to be a good remedy for "female troubles."

*Pedicularis canadensis*
Orobanchaceae

# wood betony

**HABITAT** Dry to moist prairies, savannas, open woodlands, on steep slopes

**BLOOM** Spring, summer

**DESCRIPTION** Erect perennial herb, unbranched, hairy, colonial, to 15 in. tall

**FLOWER** Yellow, burnt orange, or reddish purple, to 1 in. long, tubular, two-lipped, the upper curved and hoodlike, the lower three-lobed, on a dense spike

**LEAF** Mostly basal, lance-shaped, lobed and toothed, crinkled, fernlike, alternate, to 6 in. long and 2 in. wide

**FRUIT** Dry, smooth capsule

Also known as lousewort for the unfounded belief that livestock eating the plant would become infested with lice. The plant is partially parasitic, drawing nutrition from roots of other plants but not totally relying on them. Reported as an Ojibwe "love charm," its use is said to turn quarrelsome couples into lovers.

*Oxalis stricta*
Oxalidaceae

# yellow wood sorrel

**HABITAT** Woods, fields, roadsides, gardens, disturbed sites

**BLOOM** Summer, fall

**DESCRIPTION** Erect to ascending annual or perennial herb, branched, with spreading hairs, to 12 in. tall

**FLOWER** Yellow, five-petaled, to ½ in. wide, in clusters of one to three at or above the leaves

**LEAF** Alternate, cloverlike, with three heart-shaped leaflets, broadest at the tip and notched, to ¾ in. long and slightly wider

**FRUIT** Cylindrical capsule on upright stalks, explodes forcefully to disperse seed

Also known as sour grass, *Oxalis* plants contain oxalic acid, which gives them a sour taste. In large quantities, oxalic acid can be toxic. Each night, the leaflets fold, causing their blades to assume a more vertical position. It has been suggested that this protects the wood sorrel by causing animals that eat it to be more visible to predators.

*Linaria vulgaris*
Plantaginaceae

# butter and eggs

**HABITAT** Roadsides, fields, fencerows, disturbed sites, especially with dry soil

**BLOOM** Summer, fall

**DESCRIPTION** Erect perennial herb, unbranched, hairless, colonial, to 2 ft. tall

**FLOWER** Yellow and orange, to 1 in. long, tubular flower faces upward with two yellow upper and three lower lobes and an orangish "bump" in the center, plus a long downward spur, several in a cluster atop the stem

**LEAF** Numerous, linear, entire, alternate, hairless, to 3 in. long and ¼ in. wide

**FRUIT** Dry, rounded capsule

This snapdragon-like plant is an invasive weed introduced from Eurasia. The common name refers to the flower color, the yellow portion being the "butter" and the orange center the "egg." Also known as yellow toadflax for the similarity of flower color and leaf shape to flax (*Linum* species). "Toad" may refer to the broad mouth of the flower. Used for making yellow dye.

*Caltha palustris*
Ranunculaceae

# marsh marigold

**HABITAT** Seepage wetlands such as fens, seep springs, marshes, springs

**BLOOM** Spring

**DESCRIPTION** Erect to sprawling, hairless perennial, clumped, branched above, to 2 ft. tall

**FLOWER** Yellow, with five or six petal-like sepals, to 2 in. wide, in branched clusters atop the stem

**LEAF** Rounded to kidney-shaped, alternate on stem and basal, shallowly toothed, to 6 in. long

**FRUIT** Pod curves out and downward at maturity, resembles a jester's hat

Marsh marigold is one of the earliest blooming wildflowers in wetland habitats. Contrary to its common name, it is not a marigold or related. It is a member of the buttercup family, and its flower most closely resembles a buttercup (*Ranunculus* species). Also known as cowslip, it commonly grows with skunk cabbage.

*Ranunculus hispidus*
Ranunculaceae

# hispid buttercup

**HABITAT** Muddy floodplain forests, marshes, swamps, wet fields, meadows, thickets, ditches, riverbanks

**BLOOM** Spring, summer

**DESCRIPTION** Erect to trailing perennial herb, stems arching and some rooting at the tips, forming colonies, hairy, to 2 ft. tall

**FLOWER** Yellow, five-petaled, rounded, shiny, to 1¼ in. wide, one per stalk from upper leaf axils

**LEAF** Compound with three leaflets, each lobed or toothed, variously hairy, oval in outline, alternate, to 5 in. long and wide

**FRUIT** Flat nutlet with small beak, several packed in a cluster on a central receptacle

Generally, most buttercup plants like "wet feet," living in the same sorts of places that frogs prefer. Thus, what better genus name than *Ranunculus*, which, in Latin, means "little frog." There are two varieties of hispid buttercup in the state, the differences based mostly on characters of the flowers.

*Verbascum thapsus*
Scrophulariaceae

# common mullein





**HABITAT** Various disturbed sites, roadsides, railroads, fields, waste areas, ditches, pastures

**BLOOM** Summer

**DESCRIPTION** Erect biennial herb, unbranched (except sometime in flowering stalk), hairy, to 6 ft. tall

**FLOWER** Yellow, to 1¼ in. wide, short-tubular, five-lobed, the lower three slightly longer, numerous, in tall, spike-like cluster to 2 ft. long

**LEAF** Oblong, unlobed, blunt-tipped, basal and alternate and clasping stem, densely hairy and flannel-like, to 12 in. or more long

**FRUIT** Round capsule, each reported to contain 500–800 seeds (with up to 240,000 seeds per plant!)

The fuzzy, light green leaves of mullein are thick and soft to the touch, like flannel or felt, and have been used for shoe inserts. Introduced from Eurasia. It is said that early Greeks and Romans coated the dried flowering stalks with fat and burned them as torches. When branched, some flowering stalks resemble the shape of a saguaro cactus.

*Viola pubescens*
Violaceae

# downy yellow violet





**HABITAT** Moist to dryish forests, ravines, slopes, stream terraces, thickets

**BLOOM** Spring

**DESCRIPTION** Erect perennial herb, unbranched, hairy or not, to 15 in. tall

**FLOWER** Yellow, to ¾ in. wide, five-petaled, with two upper petals, two side petals with hairy bases, one lower petal with short spur at base, all (most prominently the lower petal) with purplish veins, solitary on stalks from leaf axils

**LEAF** Heart-shaped, one or more basal and one to four alternately on stem, toothed, hairy or not, to 4 in. long and almost as wide

**FRUIT** Cylindrical capsule, hairless or wooly, with numerous seeds

Two versions of this violet are treated as varieties or separate species. One is mostly hairless with three or more stem leaves and one or more basal leaves (known as *V. eriocarpa*, smooth yellow violet, by some); the other is hairy with two stem leaves and no or one basal leaf.

# Green Flowers

*Arisaema triphyllum*
Araceae

# jack-in-the-pulpit

**HABITAT** Rich, moist forests

**BLOOM** Spring

**DESCRIPTION** Erect, hairless perennial herb to 2½ ft. tall with a hooded, vase-shaped tube (spathe), to 4 in. tall, containing a single clublike structure (spadix)

**FLOWER** Tiny, petalless, numerous at base of club, the undersurface of the tube's flap either green or with dark purple stripes

**LEAF** One or two, divided into three leaflets to 8 in. long and 3 in. wide, with prominent veins

**FRUIT** Berries, several in a cluster, brilliantly red when ripe

A single jack-in-the-pulpit plant has the unusual trait of producing in any given year either male or female flowers, or both. It belongs to a mostly tropical family that includes several popular house plants such as heartleaf philodendron and pothos, but also a few native Wisconsin wildflowers such as skunk cabbage and wild calla.

*Ambrosia artemisiifolia*
Asteraceae

# common ragweed

**HABITAT** Fields, cropland, roadsides, disturbed sites

**BLOOM** Summer, fall

**DESCRIPTION** Erect, branched annual herb, variously hairy, to 3 ft. tall

**FLOWER** Yellowish green, tiny, in heads about ⅛ in. wide, male and female flowers occur separately

**LEAF** Pairs opposite on lower stem, upper leaves commonly alternate, deeply dissected into narrow lobes, to 6 in. long, 4 in. wide at base

**FRUIT** Tiny round bur enclosing seed

Ragweed is wind-pollinated and its airborne pollen is often a cause of hay fever for many people. Although the plant is much maligned, its seeds are important food for wildlife, especially for cardinals, chickadees, and titmice. Leaves of the similar but less common western ragweed (*A. psilostachya*) are typically thicker, hairier, and less divided.

*Polygonatum biflorum*
Convallariaceae

# smooth Solomon's seal





**HABITAT** Rich, moist forests, dryish woods, thickets

**BLOOM** Spring, summer

**DESCRIPTION** Erect to arching, unbranched perennial, hairless, to 5 ft. tall (but usually much shorter)

**FLOWER** Greenish white, tubular, with five slightly flaring lobes at tip, to ¾ in. long, two to five flowers in groups dangling from leaf axils

**LEAF** Oval, positioned somewhat on the same plane and clasping at base, hairless, alternate, numerous parallel veins visible, to 6 in. long and half as wide

**FRUIT** Berry to ⅜ in. wide, deep blue when ripe

*Polygonatum* is from the Greek *polys* (many) and *gonu* (bended knee), for the numerous joints of the rhizome. A popular explanation for its common name refers to the rhizome's leaf scars, like wax seals impressed by King Solomon's signet ring. Also in Wisconsin is hairy Solomon's seal (*P. pubescens*), with finely hairy veins on the undersurfaces of the leaf blades.

*Laportea canadensis*
Urticaceae

# wood nettle

**HABITAT** Floodplain forests, streambanks, ravines

**BLOOM** Summer

**DESCRIPTION** Erect perennial herb with a zigzagging stem covered with copious stinging hairs, colonial, to 3 ft. tall

**FLOWER** Green, about ⅛ in. wide, petals lacking, male and female separate—males on spreading branches from lower leaf axils, females from the upper axils

**LEAF** Broadly oval with small teeth along margin, alternate, to 8 in. long and 5 in. wide

**FRUIT** Shiny, flat, black nutlet

Skin contact with the stinging hairs of wood nettle brings about an immediate painful reaction. Some claim relief by application of the juicy sap of jewelweed/touch-me-not (*Impatiens* species). Fortunately, the pain is usually not long-lasting, whether treated or not. The Ojibwe have made tea from it for its diuretic properties, aiming to cure various urinary ailments.

# Blue to Violet Flowers

*Cichorium intybus*
Asteraceae

# chicory

**HABITAT** Roadsides, waste places, fields, fencerows, disturbed sites

**BLOOM** Summer

**DESCRIPTION** Erect or ascending perennial herb, branched, mostly hairless on upper stem, hairy on lower stem, milky sap, to 3 ft. tall

**FLOWER** Blue, with up to 30 ray flowers, disk flowers absent, crowded in a composite head, at tips of spreading spike-like branches, to 2 in. wide

**LEAF** Basal, also alternate on stem, with deeply triangular lobes and irregular teeth, clasping, broadest toward tip, resembles dandelion leaf, to 12 in. long and 3 in. wide

**FRUIT** Dark, five-angled nutlet

Introduced from Europe, this roadside weed has flowers that open early in the morning and usually close by midday, or sometimes later if cool and cloudy. Its roots are used for tea or added to coffee or other beverages. Its foliage is edible; some well-known domesticated chicories are radicchio and Belgian endive.

*Symphyotrichum novae-angliae*
Asteraceae

# New England aster

**HABITAT** Prairies, fields, meadows, lakesides, roadsides, fens, thickets

**BLOOM** Summer, fall

**DESCRIPTION** Erect perennial herb, branched above, hairy, to 5 ft. tall

**FLOWER** Purple, to 1½ in. wide, composed of an outer ring of up to 50 ray flowers and a similar number of yellow, tubular disk flowers; flower stalks and heads glandular, in a spreading cluster atop stem

**LEAF** Lance-shaped, unlobed, mostly toothless, hairy, base clasping, alternate, to 4 in. long and 1 in. wide

**FRUIT** Brown, narrowly wedge-shaped nutlet (seed within), tuft of hair at one end

The showy New England aster is one of the most cultivated of the native asters, with more than 50 cultivars available in the nursery trade. Another blue-flowered aster is swamp aster (*S. puniceum*). Its leaves usually have some teeth and the flower heads lack sticky glands. As its common name implies, it prefers wet environments.

*Vernonia fasciculata*
Asteraceae

# prairie ironweed





**HABITAT** Wet to moist prairies, marshes, pastures, meadows, lake borders

**BLOOM** Summer, fall

**DESCRIPTION** Erect perennial herb, unbranched, hairless, to 5 ft. tall

**FLOWER** Purple, with up to 30 disk flowers, ray flowers absent, all crowded in numerous composite heads to ¾ in. wide, clustered in a somewhat flat-topped arrangement

**LEAF** Narrowly lance-shaped with sharply toothed margins, the teeth white-tipped, undersurface mostly hairless with scattered shiny resinous pits, to 6 in. long and 1½ in. wide, alternate

**FRUIT** One-seeded nutlets with a plume of brownish orange bristles

Some say ironweed is so named because of its tough stem, but others claim the grouping of its ripe nutlets bearing brownish orange plumes suggest rusted iron. Butterflies seek out this plant, especially in late summer and fall when fewer wildflowers are in bloom. Some members of the genus show promise as a source of antitumor agents.

*Hydrophyllum virginianum*
Boraginaceae

# Virginia waterleaf

**HABITAT** Moist upland and lowland forests, ravines, floodplains, thickets

**BLOOM** Spring

**DESCRIPTION** Erect perennial herb, hairless or short-hairy, to 2½ ft. tall, rhizomatous, forming colonies

**FLOWER** Lavender to white, bell-shaped, five-lobed, to ½ in. wide, several in rounded clusters atop stem

**LEAF** Alternate, deeply lobed or compound, slightly longer than wide with three to five lobes, margins toothed, surface often with silvery white splotches, especially on the early developed leaves, to 6 in. long

**FRUIT** Rounded capsule with one to three seeds

The common and genus names refer to the silvery white splotches on some leaves. In Greek, *hydrophyllum* literally means "water leaf"—*hydor* (water) and *phyllon* (leaf). Several sources state that the leaves can be cooked as a pot herb. One of the species' common names is Shawnee salad. The Ojibwe have used the root to feed ponies to fatten them rapidly.

*Campanula rotundifolia*
Campanulaceae

# harebell

**HABITAT** Open woodlands, bluffs, rock outcrops, prairies

**BLOOM** Summer, fall

**DESCRIPTION** Erect perennial herb, slender, mostly hairless and unbranched, milky sap, to 2 ft. tall

**FLOWER** Blue, occasionally white or pink, bell- or cup-shaped, five-lobed, about ¾ in. long, nodding

**LEAF** Basal leaves rounded, often somewhat heart-shaped, usually absent during flowering; stem leaves linear and alternate, mostly hairless

**FRUIT** Capsule nodding, up to ⅓ in. long, opening by pores near base where seeds are released

Also known as bluebell, this wildflower occurs throughout the Northern Hemisphere. It is the national flower of Sweden. In North America, it occurs from coast to coast, mostly in northern latitudes and in the western mountains as far south as northern Mexico. The Ojibwe name for harebell is *adota'gons* (little bell), believed to help alleviate lung troubles.

*Lobelia siphilitica*
Campanulaceae

# great blue lobelia

**HABITAT** Floodplains, wet prairies, meadows, shores, streambanks, ditches

**BLOOM** Summer, fall

**DESCRIPTION** Erect perennial herb, unbranched, stem variously hairy and ribbed, to 3 ft. tall

**FLOWER** Blue, tubular, five-lobed, three lobes lowermost, commonly whitish near the throat, to 1 in. wide, numerous atop tall, cylindrical stalk

**LEAF** Elliptic to lance-shaped with small, shallow teeth, alternate, to 6 in. long and 2½ in. wide

**FRUIT** Capsule, round, ribbed

As indicated by its species epithet, great blue lobelia was formerly considered a treatment for syphilis. For some Indigenous tribes, it was a "love medicine," whereby roots eaten by a couple would prevent separation and renew love. Indian tobacco (*L. inflata*) and pale-spiked lobelia (*L. spicata*) have smaller flowers that are bluish (or white). Indian tobacco has been smoked by the Ho Chunk for ceremonies and everyday use.

*Tradescantia ohiensis*
Commelinaceae

# Ohio spiderwort

**HABITAT** Dry to moist prairies, savannas, roadsides

**BLOOM** Spring, summer

**DESCRIPTION** Erect perennial herb, grayish green, upwardly branching, mostly hairless, mucilaginous sap, to 3½ ft. tall

**FLOWER** Purple, to about 1½ in. wide, with three rounded petals and three smaller green sepals mostly hairless, with six hairy stalked stamens, clustered atop the stem, each flower fresh for only one day

**LEAF** Linear, folded into "V" lengthwise with a sheathing base, alternate, hairless or with scattered hairs, to 15 in. long and ¾ in. wide

**FRUIT** Capsule, somewhat egg-shaped, holds three to six seeds

The name spiderwort may be derived from the long, thread-like hairs spreading from the stamen filaments that present a spidery look. Another possibility alludes to the cobwebby appearance of dried secretions from its cut stem. Some think it is from its long, narrow leaves that arch downward, appearing like spider legs. It is also purported to be an antidote to spider bites.

*Amorpha canescens*
Fabaceae

# leadplant

**HABITAT** Dry to moist prairies and savannas, woodlands

**BLOOM** Summer

**DESCRIPTION** Erect, hairy, multi-stemmed shrub to 3½ ft. tall

**FLOWER** Purple, consisting of a single petal ¼ in. long, numerous on mostly upright spikes to 6 in. long, normally several spreading at the tips of branches

**LEAF** Pinnately compound, alternate, to 12 in. long, hairy, with up to 51 grayish leaflets

**FRUIT** Tiny pod containing a single seed

In the wild, the presence of leadplant is usually indicative of a high-quality natural area. Its common name refers to its overall grayish appearance, thought previously by some to indicate the presence of lead ore (which is untrue). Very ornamental and worthy of cultivation, it attracts a variety of pollinators.

*Dalea purpurea*
Fabaceae

# purple prairie clover

**HABITAT** Dry to moist prairies in various soil types, savannas, open woodlands

**BLOOM** Summer

**DESCRIPTION** Slender, erect perennial, hairless, to 3 ft. tall

**FLOWER** Purple, five-petaled, to ¼ in. wide, numerous, arranged in a ring on a cylindrical thimble-like head, blooming first at bottom, progressing upward

**LEAF** Pinnately compound, alternate, with three to seven linear leaflets, dark green and may be somewhat shiny, to 5 in. long

**FRUIT** Tiny pod, densely hairy at apex, with one or two seeds

This plant is beneficial to a variety of insects that are attracted to its pollen and nectar, including plasterer bees, miner bees, bumblebees, butterflies, and beetles. Called *kepiá ekié shkik* (thimble top) by the Meskwaki, its root has been made into a tea as a cure for measles. Formerly known botanically as *Petalostemum purpureum*.

*Desmodium canadense*
Fabaceae

# showy tick-trefoil

**HABITAT** Prairies, fields, savannas, woodlands, thickets

**BLOOM** Summer, fall

**DESCRIPTION** Erect perennial herb, hairy, to 6 ft. tall

**FLOWER** Pinkish purple, with typical pea family shape, to ½ in. wide, the larger, uppermost banner petal with two lighter colored spots outlined in purple (nectar guides) at its base, numerous in clusters atop plant

**LEAF** Compound, alternate with three leaflets, each to about 3 in. long, hairy

**FRUIT** Flat, segmented pealike pod with hooked hairs

Anyone walking in the fall though vegetation that includes this plant will likely become covered with its fruit. It's one of several "sticktights" that cling to clothing, shoelaces, and hair. Several wildflowers use this method of seed dispersal, which enables them to spread and establish populations in new locations.

*Lathyrus venosus*
Fabaceae

# veiny pea

**HABITAT** Dry, open woodlands, savannas, prairies, thickets

**BLOOM** Summer

**DESCRIPTION** Sprawling, climbing perennial herb, stem four-angled and mostly hairless, often in colonies, to 3 ft. tall

**FLOWER** Blue to purple, typical pea family shape, to ¾ in. long, two-toned, the uppermost petal darker than the others and with dark veins, in dense clusters of up to 20 flowers

**LEAF** Pinnately compound with up to 12 oval leaflets, to 8 in. long, each leaf tip bearing a coiled tendril instead of a leaflet

**FRUIT** Narrow, flat pod, to 3 in. long

Veiny pea is the most widespread of the upland growing wild peas in Wisconsin. Cream pea (*L. ochroleucus*) has whitish flowers and marsh pea (*L. palustris*) has fewer leaflets per leaf and occurs in wet habitats. *Lathyrus* may be derived from *la* (very) and *thuros* (passionate). The original named plant of the genus, from Europe, was thought to be an aphrodisiac.

*Lupinus perennis*
Fabaceae

# wild lupine





**HABITAT** Sand prairies, barrens, savannas, dunes

**BLOOM** Spring, summer

**DESCRIPTION** Erect perennial herb, unbranched, hairy, often colonial, to 2 ft. tall

**FLOWER** Blue-violet, typical pea family shape, to ¾ in. long, five-petaled, the upper one broad, commonly two-toned in color (blue and white), numerous on an erect stalk to 10 in. long

**LEAF** Long-stalked, alternate, palmately compound with 7 to 11 leaflets, each to 2 in. long, slightly broadest toward the tip, hairy

**FRUIT** Bean-shaped pod, hairy, to 2 in. long

Caterpillars of the endangered Karner blue butterfly (*Plebejus samuelis*) feed only on the leaves of wild lupine. Wisconsin appears to have a greater number of Karner blues than all other states combined. The nonnative large-leaved lupine (*L. polyphyllus)* typically has up to 15 leaflets that can approach 4 in. long.

*Gentiana andrewsii*
Gentianaceae

# bottle gentian





**HABITAT** Wet meadows, prairies, marshes, shores, open woodlands

**BLOOM** Late summer, fall

**DESCRIPTION** Erect to somewhat lax, usually unbranched perennial herb, essentially hairless, to 2½ ft. tall

**FLOWER** Blue, tubular, with five vertically oriented floral lobes that overlap, closed except for a tiny opening at the tip, to 1½ in. long, somewhat bottle-shaped, several clustered atop stem

**LEAF** Broadly lance-shaped, glossy, unlobed, opposite, to 4 in. long and about as wide

**FRUIT** Capsule with numerous papery winged seeds

Bottle gentian, also known as closed gentian, is one of last wildflowers of the year to bloom. Overlapping floral lobes present a challenge to potential pollinators, but large insects such as bumblebees are up to the task, strong enough to force their way through the top of the flower. Flowers can be pinkish or white, but blue is the most common color.

*Geranium maculatum*
Geraniaceae

# wild geranium





**HABITAT** Moist forests, forest edges, stream terraces, thickets

**BLOOM** Spring

**DESCRIPTION** Erect perennial herb, hairy, unbranched, usually clumping, to 2½ ft. tall

**FLOWER** Pinkish purple, with five broadly rounded petals, to 1½ in. wide, in loose clusters atop stem

**LEAF** Basal leaves and oppositely attached pairs on stem palmately divided, these further divided into broad teeth, to 5 in. long and about as wide

**FRUIT** Narrow, beaklike; when mature, outer covering splits into five separate strips that curl upward from the base, slinging seeds some distance away

Unlike commercial potted geraniums, our wild geranium is the real deal. Potted selections are of the genus *Pelargonium*, not *Geranium*. One common name for the group is cranesbill, a name referencing the long, narrow shape of the fruit, suggesting the beak of a crane. The genus name *Geranium* is from the Greek *geranos*, which means "crane."

*Iris versicolor*
Iridaceae

# northern blue flag

**HABITAT** Marshes, wet meadows, swamps, bogs, lake shores, streams, ditches

**BLOOM** Summer

**DESCRIPTION** Erect perennial herb, hairless, colonial from horizontal rhizomes

**FLOWER** Blue to violet, to 4 in. wide, parts in threes, with three petal-like outer sepals that curve downward, each bearing a greenish yellow splotch near its base, and three inner upright petals, the stalked flowers usually reach above the leaf tips

**LEAF** Stiff or arching, swordlike blades to 3 ft. tall, growing from horizontal rhizomes

**FRUIT** Three-angled, cylindrical capsule

This plant is named after Iris, the Greek goddess of rainbows. The blue flag common name likely refers to its leaves that flutter like flags in the wind. Flowers of southern blue flag (*I. virginica*) occur below the leaf tips and have sepals with a brighter yellow splotch. The Ojibwe carried a piece of blue flag, believing its scent would scare away snakes.

*Prunella vulgaris*
Lamiaceae

# self-heal

**HABITAT** Open woodlands, trails, roadsides, clearings, fields, meadows, disturbed sites

**BLOOM** Summer

**DESCRIPTION** Erect or prostrate perennial herb, mostly unbranched, hairy or not, stems four-angled, to 18 in. tall

**FLOWER** Bluish purple, ½ in. long, tubular, two-lipped, the upper one purplish and hoodlike, the lower lip three-lobed with the lowermost lobe often white, several in a dense cylindrical spike atop the stem

**LEAF** Lance-shaped to elliptic, margins may have a few shallow teeth, to 3 in. long and about half as wide, opposite

**FRUIT** One-seeded nutlets

Two varieties of self-heal exist, one or perhaps neither native to North America. It has a long history of use as folk medicine, believed to cure an assortment of ailments. Although it is in the mint family, it lacks a minty scent. The Ojibwe have made tea from its roots, said to sharpen their powers of observation for hunting.

*Scutellaria galericulata*
Lamiaceae

# marsh skullcap

**HABITAT** Marshes, swamps, wet meadows, fens, bogs, thickets, shores, ditches

**BLOOM** Summer

**DESCRIPTION** Sprawling, weak-stemmed perennial, stems square with hairs on angles, branched, to 2½ ft. tall

**FLOWER** Blue with lightly spotted white throat, tubular, to 1 in. long, two-lipped, upper forming a hood over the lower flattened lip, finely hairy, occurring singly from axils of leaves facing in same direction and appearing as pairs

**LEAF** Lance-shaped, unlobed with blunt teeth along margins, deeply veined, short stalked, progressively smaller upward, to 2¼ in. long and ¾ in. wide

**FRUIT** Rounded nutlets in a fruiting structure that resembles an old-fashioned tractor seat

Marsh skullcap occurs naturally in North America, Europe, and Asia. Of our Wisconsin species, it is most like mad-dog skullcap (*S. lateriflora*), whose flowers are almost half the size and lack spots on the lower lip. Both occur in wet habitats. The Ojibwe have used marsh skullcap to treat heart troubles.

*Lythrum salicaria*
Lythraceae

# purple loosestrife

**HABITAT** Open wetlands, marshes, meadows, shores, streams, ditches

**BLOOM** Summer, fall

**DESCRIPTION** Erect perennial herb, branched (even bushy), square-stemmed, hairy, to 6 ft. tall

**FLOWER** Purplish, to 1 in. wide, six-petaled, often with a wrinkled appearance, in dense groups on tall, narrow spikes atop the stems

**LEAF** Lance-shaped, variously hairy, opposite and somewhat clasping the stem, or some alternate or whorled, to 4 in. long and 1 in. wide

**FRUIT** Dry, egg-shaped capsule to ⅛ in. long

Despite their common names, purple loosestrife and tufted loosestrife are not related. Purple loosestrife is Eurasian in origin and extremely invasive. It negatively impacts native species and can displace them. Efforts have been made to reduce its numbers using insects from its homeland that specialize in feeding on it. So far it appears to be working.

**fireweed**
(*Erechtites hieraciifolius*)

*Chamaenerion angustifolium*
Onagraceae

# fireweed

**HABITAT** Moist clearings, roadsides, ditches, recently burned or disturbed sites

**BLOOM** Summer

**DESCRIPTION** Erect perennial herb to 7 ft. tall, colonial by rhizomes

**FLOWER** Purplish pink, four-petaled, oval-shaped, to 1 in. wide, in long cluster at stem tip

**LEAF** Narrow, spear-shaped, edges wavy or slightly toothed, alternate, to 8 in. long

**FRUIT** Slender, four-angled capsule, splits like banana peel revealing seeds, each bearing a long plume of hairs at tip

Fireweed is so named because of its common occurrence in recently burned habitats. At a distance, it may be confused with the similar looking purple loosestrife (*Lythrum salicaria*), but flowers of that highly invasive, nonnative weed of wetlands are six-petaled. Another plant called fireweed is *Erechtites hieraciifolius* (see inset), a member of the aster family with a head of tiny white flowers.

*Platanthera psycodes*
Orchidaceae

# lesser purple fringed orchid

**HABITAT** Seepage wetlands, swamps, marshes, streambanks

**BLOOM** Summer

**DESCRIPTION** Erect perennial herb, unbranched, hairless, to 3 ft. tall

**FLOWER** Purple, fan-shaped, to ¾ in. long, with three petal-like sepals and three petals, one of which is larger and lowermost with three deeply fringed lobes and a long spur at base, clustered on a stalk atop the stem

**LEAF** Lance-shaped, unlobed, hairless, alternate, to 7 in. long and 3 in. wide

**FRUIT** Capsule with numerous dustlike seeds

There are nearly 50 wild native orchids in Wisconsin, and this is one of 11 species of rein orchids (*Platanthera* species) occurring here. They are collectively called rein orchids because they have nectar spurs that fancifully resemble the reins (straps) of a horse's bridle. This orchid is known throughout the state but is most commonly found in northern parts.

*Mimulus ringens*
Phrymaceae

# Allegheny monkeyflower

**HABITAT** Marshes, fens, shores, swamps, streams, ditches

**BLOOM** Summer

**DESCRIPTION** Erect perennial herb, square-stemmed, hairless, to 3 ft. tall

**FLOWER** Blue with a yellowish center, to 1 in. long, two-lipped with wavy edges, the upper two-lobed, the lower three-lobed and widely spreading, on 1–2 in. stalks emerging from leaf axils

**LEAF** Lance-shaped with small teeth on margins, opposite, clasping at base, to 4 in. long and 1 in. wide

**FRUIT** Elliptic-shaped capsules

Known as monkeyflower perhaps because of the flower's fanciful resemblance to an expressive, "smiling" monkey or someone monkeying (or clowning) around. Such a look is also reflected in its genus name *Mimulus*, from *mimus*, for "clown." Its large, showy flowers are frequented by bumblebees, and the species is often used in pollination research.

*Pontederia cordata*
Pontederiaceae

# pickerelweed





**HABITAT** Shallow margins of lakes, ponds, bogs, marshes, swamps, rivers

**BLOOM** Summer, fall

**DESCRIPTION** Erect perennial herb, unbranched, hairless (except in flower cluster), colonial, to 3 ft. tall

**FLOWER** Blue, to ½ in. wide, two-lipped, with six spreading lobes, the uppermost with one or two yellow spots, several in a terminal spike atop the main stalk

**LEAF** Mostly basal, triangular with heart-shaped base, unlobed, toothless, shiny, noticeably parallel-veined, long-stalked and sheathing at base, to 10 in. long and about half as wide

**FRUIT** Bladderlike, housing a single seed

Pickerelweed is likely named after the pickerel, a narrowly shaped fish that, when young, may occupy the shallow water where pickerelweed grows. The pickerelweed shortface bee (*Dufourea novaeangliae*) and pickerelweed long-horned bee (*Melissodes apicata*) pollinate pickerelweed and feed almost exclusively on its pollen. Pickerelweed is wide-ranging, occurring from Canada south as far as northern Argentina.

*Anemone patens*
Ranunculaceae

# American pasqueflower





**HABITAT** Dry prairies, especially on substrates of sand or gravel

**BLOOM** Spring

**DESCRIPTION** Erect, unbranched perennial, hairy, to 15 in. tall

**FLOWER** Purple to white, to 2½ in. wide, occurring singly atop stem, five to seven sepals are petal-like

**LEAF** Deeply divided into narrow lobes, basal leaves stalked, stem leaves unstalked, barely developed when flowering occurs

**FRUIT** Seedlike fruits harbor threadlike extensions abundantly covered with hairs, giving a feathery look

In the prairie, there is perhaps no better harbinger of spring than pasqueflower. Depending on the part of the state, some of these beauties arise from the recently frozen earth and bloom in mid-April, others possibly earlier where exposed to direct rays of the sun, such as south-facing hillsides. *Pulsatilla patens* is another scientific name for this species.

*Verbena hastata*
Verbenaceae

# blue vervain

**HABITAT** Wet prairies, meadows, marshes, shores, ditches

**BLOOM** Summer, fall

**DESCRIPTION** Erect perennial herb, unbranched except in flower clusters, angled, somewhat hairy, to 5 ft. tall

**FLOWER** Blue to pinkish purple, to ¼ in. wide, tubular with five spreading lobes, numerous on dense spikes to 5 in. long at tips of branches atop the plant, somewhat candelabra-like

**LEAF** Narrow, lance-shaped, sharply toothed with deep veins on the blade surface, opposite, often with smaller leaves in axils of main leaves, to 6 in. long and 1 in. wide

**FRUIT** Small nutlets, usually in groups of four within flower remnants

The verbena bee (*Calliopsis verbenae*) specializes in acquiring food from the flowers of vervains. Vervains also serve as host plants for feeding caterpillars of the verbena moth (*Crambodes talidiformis*). The Ho Chunk have used the roots of blue vervain for treating "female weakness."

*Viola sororia*
Violaceae

# woolly blue violet

**HABITAT** Upland forests, savannas, fields, meadows, roadsides, thickets

**BLOOM** Spring

**DESCRIPTION** Erect or ascending perennial herb, unbranched, hairy, stemless (leaves and flowering stalks come directly from the rhizome), to 6 in. tall

**FLOWER** Blue with a whitened throat, to ¾ in. wide, five-petaled, with two upper petals, two side petals with hairs at their bases, and one lower petal with a short spur at its base, all (but most prominently the lower petal) with purplish veins, solitary on stalks from rhizome

**LEAF** Heart-shaped, basal, toothed, hairy, to 3 in. long and about as wide

**FRUIT** Cylindrical capsule, hairless, with numerous seeds

Also known as Wood Violet, it is the state flower of Wisconsin. The similar common blue violet (*V. communis*) is typically hairless and more likely to be found in disturbed sites. About 20 species of violets occur naturally in Wisconsin, including some with white and yellow flowers. Fritillary butterfly caterpillars, including the magnificent regal fritillary, feed exclusively on leaves of violet species.

# Brown to Maroon Flowers

leaves

*Symplocarpus foetidus*
Araceae

# skunk cabbage

**HABITAT** Seepage swamps and springs, fens, bogs

**BLOOM** Late winter, spring

**DESCRIPTION** Erect perennial herb, unbranched, hairless, colonial, to 2 ft. tall

**FLOWER** Maroon (hood), four tepals, several packed tightly together on a club-shaped structure (spadix), housed within a teepee-shaped, leathery hood (spathe)

**LEAF** Heart-shaped or oval, basal, unlobed, toothless and hairless, to 2 ft. long and half as wide

**FRUIT** Knobby compound fruit to 4 in. long and 3 in. wide, shape suggestive of a small pineapple

This is our earliest flowering native plant, blooming in some years in late winter. Perhaps its most unusual characteristic is its flowering spadix (club), capable of generating heat considerably warmer than that of the surrounding air. This trait can even melt snow on and around its hood during blooming.

*Asarum canadense*
Aristolochiaceae

# wild ginger

**HABITAT** Rich, moist forests, ravines, stream terraces

**BLOOM** Spring

**DESCRIPTION** Low-growing perennial herb, unbranched, colonial, to 6 in. tall

**FLOWER** Maroon, tubular or urn-shaped, with three long-tapering, petal-like sepals (petals absent) often curved backward, hairy on the outside, occurring singly between paired stems at ground level

**LEAF** Heart- to kidney-shaped, sparsely hairy, paired at stem tip, from creeping stem along ground, to 3 in. long and 5 in. wide

**FRUIT** Fleshy, rounded capsule

The flower of wild ginger is hidden beneath the leaves. It is scentless and, contrary to what is often stated, likely not pollinated by carrion-seeking insects. Research indicates it is mostly self-pollinating. The common name alludes to the rhizomes that smell remarkably like that of tropical ginger (*Zingiber officinale*) used in cooking.

*Sarracenia purpurea*
**Sarraceniaceae**

# purple pitcher plant





**HABITAT** Bogs, fens, swamps

**BLOOM** Summer

**DESCRIPTION** Erect to sprawling perennial herb, essentially stemless other than flowering stalks, clumping, carnivorous, to 2 ft. tall

**FLOWER** Purple to maroon, nodding, to 2½ in. wide, five-petaled, thin and deciduous, five thick sepals persistent on umbrella-like disk

**LEAF** Basal, tubular, filled with rainwater, rimmed on one side by an erect hood, hairless on the outside, the inside covered with downward pointing hairs, to 8 in. long

**FRUIT** Round to oval capsule

Pitcher plants absorb nutrients from organisms trapped in the "pools of death" in their leaves. The usual victims are insects but can include salamanders. The pitcher plant mosquito (*Wyeomyia smithii*) lays its eggs only in pitcher plant pools. Amazingly, in our area, these mosquitos do not feed on humans! An Ojibwe name for the plant is *o' makaki' wîdass*, meaning "frog's leggings."

flowers

*Scrophularia lanceolata*
Scrophulariaceae

# early figwort





**HABITAT** Open woodlands, forest edges and clearings, thickets, roadsides, fencerows

**BLOOM** Summer

**DESCRIPTION** Erect perennial herb, square-stemmed, minutely hairy, to 6 ft. tall

**FLOWER** Reddish brown, goblet-shaped, to ⅓ in. long, two-lipped, the upper two-lobed, the lower three-lobed with two lateral lobes upright and one lower lobe bent down, interior with a yellow or green, flattened sterile stamen, stalks with tiny glandular hairs, numerous at branching stem tips

**LEAF** Lance-shaped, coarsely toothed margins, straight-edged or rounded base and pointed tip, opposite, to 8 in. long and 3 in. wide

**FRUIT** Dull brown capsule narrowed to a beak at tip

Described as one of the most prolific nectar producers in the world, the inconspicuous flowers (see inset) of figworts attract hummingbirds and many insects, even bald-faced hornets. Its common name refers to its use in treating "figs," an old name for hemorrhoids. The genus name refers to its former use to treat scrofula, infected lymph nodes of the neck.

*Typha latifolia*
Typhaceae

# broad-leaved cattail





**HABITAT** Marshes, lake and pond borders, ditches, fens, wetlands

**BLOOM** Summer

**DESCRIPTION** Erect perennial herb, unbranched, hairless, colonial, to 9 ft. tall

**FLOWER** Brownish or tan (when mature), tiny and inconspicuous, sexes separate, the tan male flowers compact on the spike directly atop the much thicker portion containing the brown female flowers, reported to be as many as two million, to 1½ in. thick

**LEAF** Strap-shaped, basal and alternate, about 1 in. wide to about as long as flowering stem

**FRUIT** Tiny nutlet with "fluff" of hairs attached to its stalk

There are two species of cattails in Wisconsin, this one and the nonnative narrow-leaved cattail (*T. angustifolia*). A hybrid between the two (*T.* ×*glauca*) is quite invasive and forms monocultures to the detriment of native vegetation. The native cattail has long been important to Indigenous people, particularly for construction materials and food.

# INTERPRETING SCIENTIFIC NAMES



A scientific name often provides information about distinctive features of a plant. Many Latin words or variations are used in the binomial names of certain species treated in this book. Consider, for example, the scientific name for red clover: *Trifolium pratense*. Even without seeing the plant, you might surmise from its genus name that has three leaves (*tri* for "three," *folium* for "leaves"), and indeed this would be correct (actually, this species has three leaflets). Also, its specific epithet *pratense* means "of meadows," which is a common habitat for the plant.

| LATIN WORD | MEANING |
|---|---|
| *albidus, alba* | white |
| *alta* | tall |
| *angusti-* | narrow |
| *-anthus* | flower |
| *arvense* | field |
| *aureum* | golden |
| *bi-* | two |
| *blanda* | white |
| *boreale* | northern |
| *canescens* | white-hairy |
| *-carpum* | fruit(ed) |

| LATIN WORD | MEANING |
|---|---|
| *-caulus* | stem |
| *cernuum* | nodding |
| *cinque* | five |
| *coccineus* | scarlet |
| *cordi-* | heart-shaped |
| *crassi-* | fleshy |
| *erigeron* | early, old man |
| *flora-* | flower(ed) |
| *-folium* | leaved |
| *galericulata* | covered with a helmet or hood |
| *glaber* | smooth |
| *hirta* | short-hairy |
| *-issima* | very (superlative) |
| *lanceolatus* | lance-leaf |
| *lati-* | broad |
| *lepto-* | slender, small |
| *lithos* | stone |
| *macro-* | large |
| *maculata* | spotted |
| *mille-* | thousands, many |
| *nuda-* | naked |
| *officinalis* | healing |
| *-oides* | like |

| LATIN WORD | MEANING |
|---|---|
| *-opsis* | head |
| *palustris* | marsh-loving |
| *parva-* | small |
| *patens* | spreading |
| *pauci-* | few |
| *-phylla* | leaf |
| *platy-* | broad |
| *poly-* | many |
| *pratensis* | meadow |
| *purpureus* | purple |
| *repens* | creeping |
| *rotundi-* | round, plump |
| *rubra* | red |
| *saligna* | willow-leaved |
| *saxifraga* | rock-breaker |
| *sepium* | growing in hedges |
| *stella* | star-shaped |
| *stricta* | stiff |
| *tri-* | three |
| *umbellate* | umbrella-shaped |
| *varians* | variable |
| *villosa* | downy |
| *vulgatum* | common |

# BIBLIOGRAPHY



Curtis, J. T. 1959. *The Vegetation of Wisconsin: An Ordination of Plant Communities*. Madison: University of Wisconsin Press.

Fernald, M. L. 1950. *Gray's Manual of Botany: A Handbook of the Flowering Plants and Ferns of the Central and Northeastern United States and Adjacent Canada*. 8th ed. New York: American Book Company.

Flora of North America Editorial Committee, eds. 1993+. *Flora of North America North of Mexico*, 20+ vols. New York and Oxford: Oxford University Press.

Gleason, H. A., and A. Cronquist. 1991. *Manual of Vascular Plants of Northeastern United States and Adjacent Canada*. 2nd ed. Bronx, NY: The New York Botanical Garden.

Homoya, M. A., and S. A. Namestnik. 2022. *Wildflowers of the Midwest*. Portland, OR: Timber Press.

Ladd. D. 1995. *Tallgrass Prairie Wildflowers: A Field Guide*. Helena, MT: Falcon Press.

Ladd, D. 2001. *North Woods Wildflowers: A Field Guide to Wildflowers of the Northeastern United States and Southeastern Canada*. Helena, MT: Falcon Press.

Smith, W. R. 2012. *Native Orchids of Minnesota*. Minneapolis: University of Minnesota Press.

Tester, J. R. 1995. *Minnesota's Natural Heritage: An Ecological Perspective*. Minneapolis: University of Minnesota Press.

Voss, E. G., and A. A. Reznicek. 2012. *Field Manual of Michigan Flora*. Ann Arbor: University of Michigan Press.

Wilhelm, G., and L. Rericha. 2017. *Flora of the Chicago Region: A Floristic and Ecological Synthesis*. Indianapolis, IN: Indiana Academy of Science.

# PHOTO AND ILLUSTRATION CREDITS

Map on page 14 illustrated by Michele Angel

Cover photos by Michael Homoya, Scott Namestnik, Darren Berendt/Shutterstock, and Randy Bjorklund/Shutterstock

Flower icon by YuningArt Studio

Adam Balzer, 182

Katy Chayka, Minnesota Wildflowers, 80, 152 (inset), 214, 310, 338

Peter Dzuik, Minnesota Wildflowers, 140, 176, 248, 308

Andrew Lane Gibson, 184

Peter Grube, 180, 318

Michael Homoya, 50 (top), 60, 78 (inset), 84, 92, 96 (inset), 104, 114, 120, 134, 146 (inset), 158, 170, 172, 174, 192, 198, 224, 230, 238, 244, 250, 252, 254, 256, 262, 264, 266, 272, 290, 306, 316 (inset), 332 (bottom)

Michael Huft, 72, 118, 124, 128, 160, 274, 320, 326

Eric Hunt, 90

Scott Namestnik, 48, 52, 54, 56, 58, 62, 64, 66, 68, 70, 74, 76, 82, 86, 94, 96, 98, 100, 106, 108, 110, 112, 116, 120 (inset),122, 126, 130, 132 (inset), 136, 138, 142, 144, 146, 148, 150, 152, 156, 162, 166, 168, 186, 188, 196, 200, 202, 206, 208, 212, 216, 218, 220, 222, 226, 228, 232, 234, 236, 240, 242, 246, 258, 260, 270, 276, 280, 282, 286, 288, 292, 294, 296, 298, 302, 304, 314, 316, 322, 328, 332 (top), 336, 340

Nathanael Pilla, 178

Corey Raimond, 50 (bottom), 150 (inset), 300, 324

Paul Rothrock, 176 (inset)

Perry Scott, 88, 102, 164, 190, 210, 312

B.S. Slaughter, 284, 334

Dan Tenaglia, 78

## Shutterstock

Alan B. Schroeder, 40–41

Brian Woolman, 38

BrittJPete, 132

Carmen Rieb, 20

Cynthia Shirk, 5

Darren Berendt, 8

Kevin Collison, 12

Mike Truchon, 42

torook, 2–3

## WEB RESOURCES



The Biota of North America Program: bonap.org

Consortium of Midwest Herbaria: midwestherbaria.org

Flora of North America: eFloras.org

Flora of Wisconsin: wisflora.herbarium.wisc.edu

Illinois Wildflowers: illinoiswildflowers.info

Michigan Flora (University of Michigan Herbarium): michiganflora.net

Minnesota Wildflowers: minnesotawildflowers.info

NatureServe network of Natural Heritage Programs: natureserve.org/natureserve-network/united-states

## STATE NATIVE PLANT SOCIETIES

Every state in the Midwest has one or more organizations that focus on native plants. Readers are encouraged to consult websites of several native plant–focused organizations.

North American Native Plant Society: nanps.org

Botanical Club of Wisconsin: sites.google.com/site/botanicalclubofwisconsin/join-bcw

Michigan Botanical Society: michiganbotanicalsociety.org

Minnesota Native Plant Society: mnnps.org

The Prairie Enthusiasts: theprairieenthusiasts.org

# INDEX

© Barbara Homoya

**Michael Homoya** began his career teaching science at the Pine Ridge Indian Reservation in South Dakota and biology at Southwestern Community College in North Carolina. For thirty-seven years, he served as botanist and plant ecologist for the Indiana Department of Natural Resources before retiring in 2019. He is a fellow and former president of the Indiana Academy of Science and a board member and former president of the Indiana Native Plant Society. His previous books include *Orchids of Indiana*; *Wildflowers and Ferns of Indiana Forests: A Field Guide*; *Wake Up, Woods* (a children's book); and *Wildflowers of the Midwest* (co-authored with Scott Namestnik). He was awarded the Distinguished Career Public Service Award from the Conservation Law Center, the Distinguished Scholar of the Year from the Indiana Academy of Science, and the Barbara J. Restle Lifetime Conservation Award from Sycamore Land Trust.